John P. Story

Elements of Elastic Strength of Guns

A text book for the use of student officers of the U.S. Artillery School

John P. Story

Elements of Elastic Strength of Guns
A text book for the use of student officers of the U.S. Artillery School

ISBN/EAN: 9783337015596

Printed in Europe, USA, Canada, Australia, Japan

Cover: Foto ©berggeist007 / pixelio.de

More available books at **www.hansebooks.com**

ELEMENTS OF
ELASTIC STRENGTH OF GUNS.

A TEXT BOOK

FOR THE USE OF STUDENT OFFICERS AT THE

U. S. ARTILLERY SCHOOL,

BY

CAPTAIN JOHN P. STORY,

Fourth Artillery, U. S. Army,

INSTRUCTOR.

ARTILLERY SCHOOL PRESS
FORT MONROE, VIRGINIA.
1894.

COPYRIGHT.
1894.

PREFACE.

This work has been prepared for use as a text book in the U. S. Artillery School at Fort Monroe, Va., and also as a book of reference for artillery officers who may be interested in the principles upon which depend the construction of the built-up steel gun.

The text is divided into three parts.

In Part I will be found the preliminary definitions and the deduction of the fundamental stress formulas. These equations, (7) and (8), in the form adopted, have the sanction of such eminent authority as Duguet of France and of Pashkievitsch of Russia; in no other form will an extending stress be always positive and a compressing stress negative.

In Part II is discussed the Elastic Strength of Guns with respect to *stress*. Advantage has been taken in this part to give a graphic illustration of the distribution of stresses in a built-up gun both in the state of rest and in the state of firing.

In Part III is discussed the Elastic Strength of Guns when the resistance is measured by *strain*, or deformation. This discussion is based upon Clavarino's formulas with the important modification in the signs and values of the numerical coefficients determined by Captain Rogers Birnie, Ordnance Department, U. S. Army.

Part III is of special interest to the American artillerist since our modern guns of home manufacture are constructed by the application of *strain* formulas; in this Part, following the practice of the Army Ordnance Department, the radial elastic strength of the several cylinders of a built-up gun are determined,

As steel is the only metal now used in gun construction a *constant* modulus of elasticity has been adopted— $E = 13.393$ tons.

Although there has been no hesitation in using any literature on gun construction available, yet special acknowledgments are due to Captain Tasker H. Bliss, Aide-de-Camp to the Major General Commanding the Army, for use made of his excellent

PREFACE.

translation, from the Russian of Colonel Pashkievitsch, Professor at the Michael Military Academy, Russia, of "The Resistance of Guns to Tangential Rupture," and also for the courtesy of a special translation, for the personal use of the writer, of a "Collection of Formulas" by the same author, which formulas were "examined and approved by the Artillery Committee".

J. P. S.

TABLE OF CONTENTS.

PART I.

1—Introduction. 2—Stress. 3—Strain. 4—Relation between stress and strain. 5—Elasticity. 6—Elastic limit. 7—Special elasticity. 8—Hooke's law. 9—Modulus of elasticity. 10—The elastic resistance of a prism. 11—Elastic strength of a tube under fluid pressure. 12—Elastic strength of a thin tube. 13—Elastic strength of a thick tube.

PART II.

STRESS.

14—The simple elastic strength of a tube. 15—Discussion of "stress" equations. 16—*First case:* $P_1 = 0$. 17—Curve of tensions, and of pressures when $P_1 = 0$. 18—*Second case:* $P_0 = 0$. 19—Curve of tangential compressions, and of pressures when $P_0 = 0$. 20—*Third case:* When P_0 and P_1 both act. 21—Tensions and pressures at rest, and of firing. Natural tensions, and pressures of firing. 22—A compound tube, and initial tensions. 23—The elastic strength of a tube with initial tensions. The nature of reinforcement. 24—The elastic resistance of a built-up gun in terms of the tangential tensions. 25—The most advantageous radius for a shrinkage surface. 26—Tangential elastic strength of a built-up gun with most advantageous radii and constant elastic limits. 27—The most suitable thickness of wall of a perfectly reinforced gun. 28—The longitudinal resistance. 29—Firing pressures. 30—Shrinkage formulas using firing pressures. 31—Illustration of the advantages of shrinkage.

PART III.

STRAIN.

32—Preliminary discussion. 33—The compound elastic strength of a tube. 34—Discussion of "strain" equations—*First case:* $P_1 = 0$. 35—*Second case:* $P_0 = 0$. 36—*Third case:* When P_0

and P_1 both act. 37—Strains at rest, and of firing. Natural strains of firing. 38—The elastic strength of a tube with initial tensions. The nature of reinforcement. 39—The elastic resistance of a built-up gun in terms of the tangential strains. 40—The most advantageous radii for shrinkage surface. 41—Tangential elastic strength of a built-up gun with most advantageous radii and constant elastic limits. 42—The most suitable thickness of wall of a perfectly reinforced gun, with constant elastic limits. 43—The longitudinal resistance. 44—Firing pressures. 45—Shrinkage formulas using firing pressures. 46—Pressures at rest. 47—Shrinkage formulas using pressures at rest. 48—Application of "strain" formulas.

ELEMENTS OF ELASTIC STRENGTH OF GUNS.

PART I.

INTRODUCTION.

1. The name Elastic Strength of Guns is given to that part of the theory of Artillery Engineering which deals with the nature and effect of stresses or strains in gun construction; its principal object is to determine the pressures which can be safely applied to pieces whose dimensions and arrangements are dependent upon the physical tests of the metal to be employed. The subject comprises experimental investigation into the properties of gun metal as to strength and elasticity, and a mathematical discussion of the stresses or strains in a gun, due to its mode of construction, due to fluid pressure resulting from explosion, or of a combination of these two.

STRESS, STRAIN, ELASTICITY.

2. STRESS is a force applied to a body which produces a definite alteration in size or shape; or it may be more accurately defined as the application of equilibrating forces to a body. The only stresses considered in gun construction are stresses of extension or of compression; an extending stress is a pull; a compressing stress is a push, and both are considered as acting normally to the surface considered. The word stress conveys the idea of an application of a force to a body, a resistance and a resulting equilibrium. If the force exceeds the resistance a separation into parts or rupture will occur.

INTENSITY OF STRESS is the number of units of force per unit of area and is usually expressed in the United States in tons weight or pounds weight per square inch.

Stress is *uniformly* distributed over a surface when each fraction of the area of surface bears a corresponding fraction of the whole stress.

3. Strain is the change of size or shape, that is *deformation*, produced by stress. If the stress is a simple longitudinal tension or extending force the strain consists, as shown by experiments, of an elongation in the direction of the stress accompanied by a contraction in all directions at right angles to the stress. If the stress is a simple compressing force, the strain consists of a contraction in the direction of the stress, and of an elongation in all directions at right angles to it. Hence if tension and the strain in the direction of the tension be regarded as positive, the lateral contractions negative; then it follows that a compressing stress and the strain in the direction of the compression will be negative and the lateral elongations positive. Therefore if the sign of the strain in any direction is positive we know the action of the stresses in that direction is extending; if the sign is negative the action of the stresses is compressing. It has been determined by experiment that the lateral strains are always of less numerical value than the direct strains, but if the stresses are within the elastic limit, they bear a constant relation to them.

Strain is measured by the change *in unit of length*; in gun construction it is usually expressed in terms of the linear inch; for any particular test specimen it is the ratio of total change in length to original length. Since this ratio is the quotient of two magnitudes of the same kind it may be considered as an abstract number.

RELATION BETWEEN STRESS AND STRAIN.

4. Beyond the elastic limit the relation of strain to stress becomes very indefinite, but as before stated the strain increases more rapidly than the stress.

Figure 1 represents a curve of *stress* and *strain* which is based upon the loads applied to a test specimen cut from a steel hoop made by the Midvale Steel Co., for the first experimental 8" B. L. steel rifle (experimental section, the data will be found on page 296: "Tests of metals" 1885). The stresses are the loads applied in the testing machine at Watertown Arsenal to a bar of steel and the resulting elongations were measured by a micrometer. In the figure the abscissas represent the applied stresses and the ordinates the corresponding tabulated elongations. The

ELEMENTS OF ELASTIC STRENGTH OF GUNS.

FIG. I.

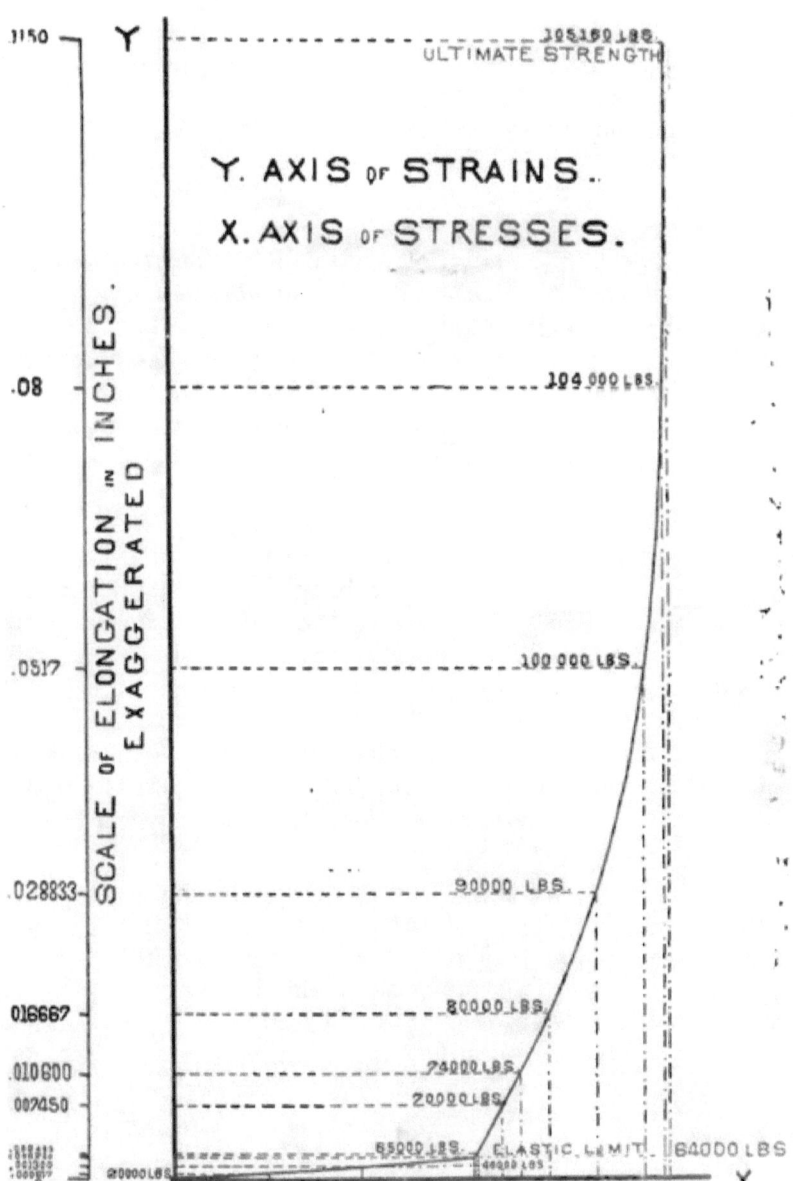

curve is seen to be practically straight (thus satisfies *Hooke's* law) from the origin of stresses to a stress of 64,000 pounds—the elastic limit—and then to bend sharply to the left; **soon** becoming nearly straight in the new direction. The figure shows that beyond the elastic limit the elongations increase much more rapidly **than do the** stresses, most **of the** elongations **being due** however to permanent strain. The **test** specimen **on which** Figure 1 is based, did not rupture until a stress of **105,160 pounds had** been applied and the resulting elongation was **more than** fifty-five times its amount at the elastic limit. The **rupturing or** breaking stress is called the *strength* or *ultimate tenacity* of the metal; *strength* is expressed in lbs. **or tons per square inch.** The **trial test** on which this curve of *stress* **and** *strain* is **based may be taken as a type** of many thousands **of such** trials made **on iron** and steel. It shows, as a rule, that in these metals **a tolerably well** defined limit exists within which the material is nearly **perfectly** elastic; the very small deviations in increments of elongation for **equal** increments of load being **more due** to a want **of** homogeneity in the test specimen than **to actual** defect **in** elasticity.

5. ELASTICITY **is** the property **a body** possesses **of** returning to its original size and shape after **the** forces, which have been applied to it, have been removed; that is, **a** body is perfectly elastic with regard **to any** applied stress, if **the** strain disappears when the stress is removed. But this restoration **to** original size and shape does not always take place; **if the** applied **stress is too** great, the strain produced in **the body will** only **partially disappear** upon the suppression of the **force,** and the bo**dy will receive** a permanent deformation.

6. ELASTIC LIMIT **is** the greatest stress which applied to a body will not produce a permanent **strain;** in practice, the stress, next **below** the one which in a testing machine will leave after removal an appreciable deformation **in** the test specimen, is usually recorded as the *elastic limit;* hence if the applied stress is less than the elastic limit, the strain which in metals is usually small in amount disappears when the stress is removed; if the applied stress exceeds the limit, the strain is in general much greater than before, but an increasing part is found when the stress is removed to consist of **a** permanent strain. In gun con-

struction the *elastic limit* is considered only with respect to extending stress and compressing stress, called respectively the *elastic limit for extension* and the *elastic limit for compression*. In this text the former is represented by θ and the latter by ρ and each limit is determined by free mechanical tests in a standard testing machine.

Experiments have shown that for mild steel the *elastic limit for extension* and the *elastic limit for compression* are practically equal; if there be a difference the greater value probably belongs to the elastic limit for compression; however this can only be determined by more extensive experiments.

On account of its application to gun construction, it is important to note that Wöhler's experiments have shown that for cast-steel the resistance to alternate stresses of extension and of compression, provided the recurring stresses were within a certain range of the elastic limit, is practically unlimited. It is however certain this metal will not, for any great length of time, resist rupture when worked to the extreme double limit of alternate stresses of extension and of compression. Wöhler found that the steel he experimented upon would bear an unlimited number of fluctuations of stresses, between the limits, —29,000 lbs. to +29,000 lbs.; 0 to +50,000 lbs.; 35,000 lbs. to 80,000 lbs.

Experiments to determine the ranges between which gun steel will bear unlimited repetition of alternating stresses would be very valuable; if the resistance of a gun were kept within such bounds, its life, except for erosion, would be practically unlimited, however frequently fired.

7. SPECIAL ELASTICITY:—Steel, iron, bronze and some other metals possess a remarkable property called *"special elasticity"* which may be stated thus, "suppose a metal subjected to a force sufficient to stretch its fibres beyond their elastic limit and to permanently elongate them; it is then placed in a new elastic condition different from the first and its limit of elasticity is greater; therefore if the body is in this new condition which we have called 'special elasticity,' it can, without further deformation, sustain forces inferior to that which deformed it, but superior to that corresponding to the original elastic limit of the body." Compressing stress as well as tensile stress develops

special elasticity. If the principle enunciated is exact, it is evident that if we subject metallic bodies to a greater stress than their limit of elasticity they must, after the withdrawal of the force, provided the body is not ruptured, have acquired a special elastic strength greater than they originally possessed. Wire possesses to a remarkable degree the property of special elasticity which is induced by the process of cold drawing. The metals which sustain the greatest elongation before rupture are as a rule susceptible of receiving the greatest increase in elastic resistance, but it should be noted that with a gain in special elastic strength there is a loss in ductility.

It appears that the property of special elasticity is only developed by "cold treatment." Experiments show that the elastic strength is not increased by an extending stress, when that stress is obtained by expanding the specimen by heat and restraining the contraction due to cooling.

The enlarging of steel or bronze unheated tubes on a mandrel seems to set up both special elasticity and a natural initial tension; that is to say there is an increase of elastic strength, and there are developed at the interior surface of the tube stresses or strains of compression, and at the exterior surface stresses or strains of extension.

8. The investigation of the strength of material within the elastic limit depends on *Hooke's* law, *i. e.*—*The strain produced by a stress is proportional to the stress producing it.* (Ut tensio sic vis.)

9. MODULUS OF ELASTICITY—If the face of a rectangular prism be subjected to a stress acting in the direction of the length of the prism, we will find a strain in the direction of the stress, and so long as the stress is within the elastic limit of the material, the stress will be proportional to the strain or the following ratio will be constant:

$$\frac{Stress}{Strain}$$

This ratio is called the Modulus of Elasticity and is usually represented by E, hence we may have:

$$E = \frac{Stress}{Strain}; \therefore \text{STRESS} = E \times Strain; \text{STRAIN} = \frac{Stress}{E} \quad (1)$$

These equations show that E is a stress and the Modulus of Elasticity may be defined as the theoretical stress, which applied to unit length will produce unit strain; this definition is readily deduced from any of the preceding equations, by substituting 1 for *strain*.

Although they are frequently confounded, it is important to note that the *modulus of elasticity* and the *elasticity* of a material are two essentially different things. As a matter of fact there is no connection between these ideas. Elasticity is defined by the entirely theoretical consideration of a perfectly elastic material, which is the property which certain substances have of returning exactly to their primitive dimensions, after the stresses, which deformed them, have been removed.

This property is entirely theoretical, since there exists no perfectly elastic body. But there are substances approaching more or less this ideal state. We know practically that elasticity may be considered as perfect in metals up to a certain limit; that is to say so long as the stress applied to a metal does not reach a certain intensity, the elastic deformation (strain) is proportional to the stress producing it, and its permanent strain is absolutely negligible; and as this property does not hold good when this stress is too great, we may consider the elasticity as measured by the elastic limit; or in other words the higher the elastic limit the more elastic is the body.

Besides this property of elasticity there is another of great importance in the selection of a gun metal, which is its *stiffness* or *rigidity*. Experiments show that within the elastic limit different metals suffer greater or less strains for a given stress, and that one metal is more rigid than another when a greater stress must be applied to it to produce an equal strain. This last property is measured by the *modulus of elasticity*, and we see the application of the preceding definition of the term as being "the *theoretical* stress which applied to unit length will produce unit strain;" supposing of course that it is possible to attain such a strain without exceeding the limit of elasticity and the tenacity of the body.

Therefore the modulus of elasticity and the elasticity of a metal are two essentially different things, and if we examine the numerical values of these two properties for different metals as determined by experiment, it will be evident that there is no bond between them. For instance all mild steel has a modulus of elasticity varying so slightly from 30,000,000 lbs. that in this text we have assumed that value as a *constant modulus* for gun steel, while the same steel, depending on the treatment it receives, may have an elastic limit varying from 30,000 lbs. to 100,000 lbs. Again, steel wire having an elastic limit of 230,000 lbs. has substantially the same modulus of elasticity as the bar steel from which it is made. Therefore it must not be said that one metal is more elastic than another, when under the same load it suffers greater strain.

The *modulus of elasticity* may be determined for any test specimen of given length, by applying a stress and measuring the corresponding strain; then apply the formula. For illustration, let the data be: Length of test bar 5 inches; area of cross section ½ square inch; load (tensile stress) 24,000 lbs.; increase of length 0.008 inches.

$$Stress = 24{,}000 \times 2 = 48{,}000,$$
$$Strain = 0.008 \div 5 = 0.0016,$$
$$E = 48{,}000 \div 0.0016 = 30{,}000{,}000.$$

In determining the modulus of elasticity it is not usual to consider stresses and corresponding strains, which are near either to the zero limit or to the elastic limit.

THE ELASTIC RESISTANCE OF A PRISM.

10. The way in which the stress produces strain differs according to the nature of the material. In this investigation the material will be considered as *isotropic*, that is perfectly elastic and having the same elastic properties in all directions, and also the following fundamental proposition, which is accepted in the theory of elasticity, will be assumed: "When several forces act simultaneously on a prism producing deformations within the elastic limit, their complete effect is obtained by taking the algebraic sum of the components."

Let us suppose a rectangular prism under the action of three rectangular extending stresses t, p and q, acting respectively in the direction of the axes X, Y and Z.

In the direction of the axis of X there will result a strain of elongation, $\frac{t}{E}$ (Article 9) due to the action of the stress t, and in and in the direction of the same axis due to the tensile stresses p and q there will result two strains of contraction, one $-\frac{mp}{E}$, the other $-\frac{mq}{E}$, in which m is a coefficient and less than 1 (Article 3).

The total strain in the direction of the axis of X will be the algebraic sum of these three separate strains and denoting this total strain by $[t]$, we will have $[t] = \frac{t}{E} - \frac{mp}{E} - \frac{mq}{E}$

Adopting a similar notation, we may in the same way deduce the total strains in the direction of the axes of Y and Z and thus have:

$$\left.\begin{aligned}[t] &= \frac{1}{E}\left(t - mp - mq\right) \\ [p] &= \frac{1}{E}\left(p - mt - mq\right) \\ [q] &= \frac{1}{E}\left(q - mt - mp\right)\end{aligned}\right\} \quad - \quad - \quad - \quad - \quad (2)$$

The first members are the symbols for the total strains in the direction of the corresponding axis, and the second members show the composition of the total strains from the separate strains when all the stresses act simultaneously.

Had the stresses all been compressing (push), the stresses and strains would all have had signs opposite to those in equations (2) but the equations would have been of the same general form. It is also evident the general form of the equations would not be changed if the stresses had been a combination of extension and of compression. To compute the proper arithmetical values from equations (2) it is necessary in substitution of the values of the stresses and strains to affect them with their proper algebraic signs.

Equations (2) show that the effect of the three principal stresses and consequently of any state of stresses on isotropic matter is

---

* The symbols $[t]$, $[p]$, and $[q]$, may be read *strain t, strain p,* and *strain q.*

Elements of elastic strength of guns. 2.

to produce three **strains**, the axes **of which** coincide **with axes of stress** and in which the principal strains are connected **with the principal stresses by** these three **equations.**

Wertheim from observations of the alterations in the **dimensions of a** body corresponding to a **definite** alteration **of its proportions** in one direction, determined the **value of** m **for iron and mild steel** as $\frac{1}{3}$. The experimental **value of** $m = \frac{1}{3}$ **is called** *Wertheim's coefficient*, **but** this value **is not** universally **admitted as exact.*** The value of $m = \frac{1}{3}$ is however **the** value adopted **by the** Ordnance Department, U. S. Army, **in** its artillery **engineering,** and substituting it in equations (2) **we have:**

$$[t] = \frac{1}{E}\left(t - \frac{p}{3} - \frac{q}{3}\right)$$
$$[p] = \frac{1}{E}\left(p - \frac{t}{3} - \frac{q}{3}\right) \qquad - \quad - \quad - \quad - \quad (3)$$
$$[q] = \frac{1}{E}\left(q - \frac{t}{3} - \frac{p}{3}\right)$$

ELASTIC STRENGTH OF **A TUBE UNDER** FLUID PRESSURE.

11. In the case of a tube when **the** internal and external pressures are different, it is clear that **the** stresses at any point of the walls may be resolved into three **at right angles to** one another; **the** first parallel **to** the axis (*longitudinal tension* q); the **second** perpendicular to the axis (*radial pressure* p); **and the third per**pendicular to the plane of the other two (*tangential* or *hoop tension* t); the third is tangential to a co-axial **cylinder passing through the** point considered.

A gun is essentially **a tube and when** powder is **exploded, the** chamber until the projectile begins to move, may be regarded as a closed cylinder subject to internal **fluid** pressure. As the pressure of **a** fluid acts equally in all directions **and** normally to the **sur**faces with which it **is** in contact, the powder pressures produced **by** explosion act radially to the curved surface of the bore and **also** at right angles **to the** end of the bore and the **base of the pro**jectile.

---

* General Virgile in his **earlier** investigations accepted the value of $m = \frac{1}{3}$; but later he adopted the value of $m = \frac{1}{4}$ as proposed by Poisson, **and this** value of $\frac{1}{4}$ is the one accepted by the French Artillery in Gun Construction.

ELEMENTS OF ELASTIC STRENGTH OF GUNS.   11

ELASTIC STRENGTH OF A THIN TUBE.

12. When a tube is subjected to internal fluid pressure, this pressure acting radially produces in the walls of the tube a stress of circumferential pull or tension; this stress acts in the direction of a tangent to a cross section and is usually called tangential tension. If as generally happens there is acting also an external fluid pressure less than the internal, then in determining the mean stress the intensity of the internal pressure must be taken as the excess of the internal over the external pressure. In an indefinitely thin tube the tangential stress is uniform throughout the thickness, and when the thickness of the tube is small as compared with the diameter, the tangential stress may be considered as uniformly distributed over the thickness.

FIG. 2.

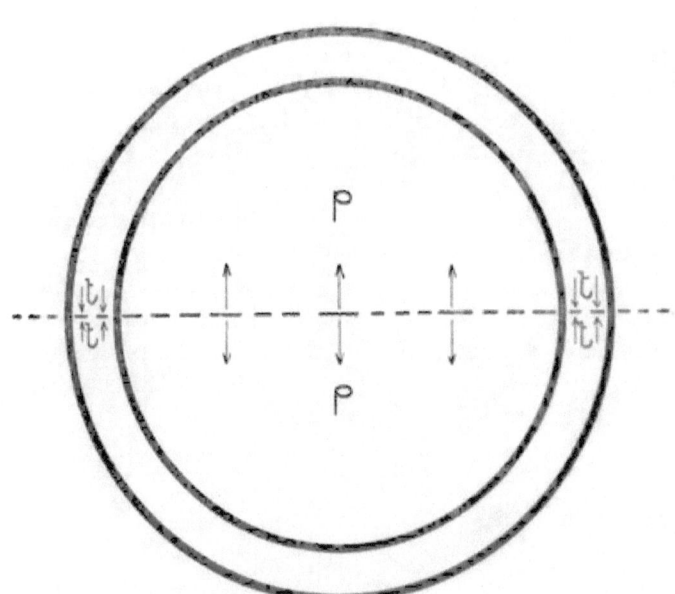

Let Figure 2 represent the cross section of a thin tube of unit length. A plane through the axis will divide the tube into two halves, and since it is a fundamental law of fluid pressure that the fluid pressure in any direction is equal to the pressure on a plane perpendicular to that direction, the tendency to produce separation along the plane is $2 R_0 p$ in which $2 R_0$ is the internal diameter, and p the intensity of fluid pressure acting at right angles to the plane considered. The resistance to separation is equivalent to the uniform tensile stress developed by the radial pressures in the walls of the tube. Let R_1 represent the exterior radius and t the mean intensity of tangential tension acting over the thickness $R_1 - R_0$ of the tube. The amount of this stress is $2(R_1 - R_0) t$. And since the two stresses are in equilibrium we have:

$$2 R_0 p = 2(R_1 - R_0) t$$
$$\therefore \quad p = \frac{t(R_1 - R_0)}{R_0} \quad - \quad - \quad - \quad (4)$$

Having a constant thickness $R_1 - R_0$ and a given elastic strength t of material, we see that as R_0 increases p diminishes or that small tubes of a definite thickness will sustain with safety an internal pressure, which tubes of larger diameter with equal thickness will not bear.

This formula should only be used when the thickness is small as compared with the diameter. If the thickness is great the tensile stress will not be uniform over the section, and the formula will then give too great a value for p.

It is evident that if the tube be subjected to external fluid pressure, that there will be developed in a direction perpendicular to the external diameter, $2 R_1$ a stress equal to $2 R_1 p'$ and the tendency to crushing will be resisted by the stress of compression t, produced in the walls of the tube by the pressure acting normally to the other surface, the developed stress acting over twice the thickness. Hence we readily deduce:

$$p' = \frac{t(R_1 - R_0)}{R_1} \quad - \quad - \quad - \quad (5)$$

A comparison of this equation with (4) shows they only differ by the diameters over which the radial pressures act, also that a tube will bear a greater internal than external pressure.

Next regarding the tube as closed at both ends its tendency to rupture over a cross section is measured by the pressure on each end, $\pi R_0^2 p$ and the resistance to separation is equivalent to the tensile stress q (longitudinal tension) called into action over the annular cross section $\pi(R_1^2-R_0^2) q$.

Hence the conditions of equilibrium are:

$$\pi R_0^2 p = \pi(R_1^2-R_0^2) q.$$

$$\therefore \qquad q = \frac{R_0^2 p}{R_1^2-R_0^2} \qquad - \quad - \quad - \quad (6)$$

ELASTIC STRENGTH OF A THICK TUBE.

13. Eqations (2) connecting stress and strain in combination with suitable equations expressing the continuity of the body and the equilibrium of each of its elements, are theoretically sufficient to determine the distribution of stress within an elastic body exposed to given forces, and in particular to determine the parts of the body exposed to the greatest stress and the magnitude of such stress.

In deducing the equations expressing the equilibrium of stress in a tube, it is necessary to consider the way the tube yields under the application of the forces to which it is subjected. The simplest way to do this is to assume that the tube remains still a tube after the pressures have been applied, or that the longitudinal tension in the walls of the tube is uniform. Nor is this an arbitrary assumption; no other apparently can be made if the ends of the tube are free and the pressures on the surfaces of the tube exactly uniform.

A homogeneous tube may be considered as composed of an infinite number of elementary cylinders infinitely thin, concentric, superposed and exerting in a state of rest no force upon one another.

Let us assume a tube, forming part of a gun, which is subjected to a uniform internal pressure; for instance the pressure of powder gases. Let there be applied to the external cylindrical surface a pressure uniformly distributed, which may be either the pressure of the atmosphere or the pressure due to hoops which have been shrunk on the tube. It follows from the assumptions just enunciated.

First.—That the particles of the tube which were situated before firing in a right section of the tube, will, at the moment of firing, change their position in such a way that, equilibrium being preserved, they will still remain in their right section and the two right sections have the same center; whence the longitudinal strain will be uniform.

Second.—That the particles of the tube which before firing were on a cylindrical surface whose axis is the axis of the tube, will continue in the moment of firing upon a cylindrical surface which has this same axis.

With these assumptions the tube will remain a tube at the moment of firing, but with a different thickness of wall and a different length from what it had before firing.

When a homogeneous tube is exposed only to an internal fluid pressure, due for example to the explosion of gunpowder, all the elementary cylinders which make up the tube exert upon each other pressures, acting radially outward, which diminish from the interior where the pressure is equal to that of the powder gas, to the exterior where it is zero. At the same time since the tube is in equilibrium, each elementary cylinder sustains a tension which equilibrates the pressure.

Let us now consider a tube which supports only an exterior radial pressure; all the elementary cylinders exert pressures upon each other, but the radial pressure diminishes from the exterior to the interior surface where it is zero.

Finally let us assume that a homogeneous tube is exposed simultaneously to two pressures, one internal, the other external. At any point of the tube there will be a resultant pressure which may, depending upon the relative values of the internal and external pressures at the point, be acting either outward or inward. At all the points of the tube the stresses will be assumed at right angles to each other as explained in Article 11.

In the interior of a tube let us assume an elementary cylinder $A' B_1 C - A B C$ of unit length with radii r and $r+dr$, Figure 3.

Let us denote by:

R_0, interior radius of tube.
R_1, exterior radius of tube.
r, radius of cylindrical surface $A' B_1 C'$.
$r+dr$, radius of cylindrical surface $A B C$.

P_0, pressure at interior surface of the tube acting radially outward.

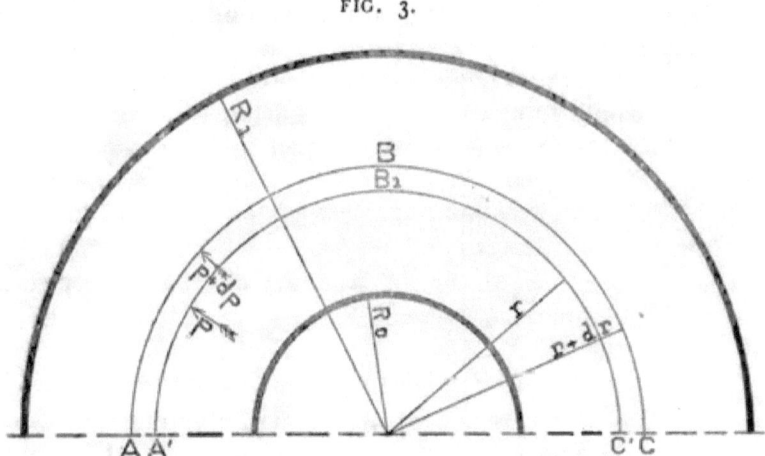

FIG. 3.

P_1, pressure at exterior surface of the tube acting radially inward.

p, radial pressure at distance r from the axis.

$p+dp$, radial pressure at distance $r+dr$ from the axis.

q, uniform stress parallel to the axis of the tube.

t, tangential tension, at the distance r from the axis due to the pressure p.

T_0, T'_0, respective values of t when $r=R_0$ and $r=R_1$.

E, modulus of elasticity of the tube.

We will obtain one equation showing the relation between p and t by taking the third of equations (3) which gives the strain in the direction of the axis of the tube and have:

$$[q] = \frac{1}{E}\left(q - \frac{t}{3} - \frac{p}{3}\right)$$

or
$$t + p = 3(q - E[q])$$

Since the stress and strain in the direction of the axis of the tube are uniform the second member is constant and we may write:

$$t + p = c_1 \qquad - \quad - \quad - \quad - \quad (7)$$

The pressure p, resulting from the combined action of P_0 and P_1, acts radially over the surface $A'B_1C'$ and develops in a direction perpendicular to $A'C'$ a pressure $2\,rp$ which is the stress tending to rupture the tube at the surface whose radius is r.

The pressure acting radially at the exterior surface ABC of the elementary cylinder will be $p+dp$, and the tendency to produce rupture at that surface will be $2(r+dr)(p+dp)$. Since the tube is in equilibrium with the stresses acting upon it, the difference of these two pressures will be equal to the resistance of the elementary cylinder $A'B_1C'—ABC$; which resistance, since t may be considered as constant through an infinitely thin wall (Article 12) is equal to $2\,t\,dr$.

Since the tendency to rupture is equilibrated by the resistance, the algebraic sum of these parallel stresses is equal to 0 and we have:

$$2\,pr - 2(r+dr)(p+dp) + 2\,t\,dr = 0$$

by reduction and omission of the term containing the second differential we obtain:

$$\frac{dr}{r} = \frac{dp}{t-p}$$

substituting for t its value from (7) we have:

$$\frac{dr}{r} = \frac{dp}{c_1 - 2p}$$

integrating

$$lr = -\frac{1}{2} l\left(c_1 - 2p\right) + l\,c$$

or

$$r^2 = \frac{c^2}{c_1 - 2p}$$

since $t = c_1 - p$ we obtain, after replacing c^2 by c_2, and reducing:

$$(t-p)\,r^2 = c_2 \qquad (8)$$

Although other equations will be deduced more convenient of application to gun construction than (7) and (8), yet these are the *fundamental* equations expressing the conditions of equilibrium of stresses in a tube.

From (7) and (8) the following remarkable properties may be enunciated:

First.—At all points the sum of the tangential tension and the radial pressure is the same.

Second.—At any point whatever the difference of the tangential tension and the radial pressure varies inversely as the square of the radius.

From formulas (7) and (8) we may deduce the condition at any point of any tube when the exterior and interior pressures, or tensions together with the tubes dimensions are given.

Applying the notation of (7) and (8) to a separate tube, or to the innermost tube of a built-up gun we shall have, since the second members are constant:

$$t + p = T_0 + P_0 = c_1 \qquad (9)$$
$$T_0 + P_0 = T'_0 + P_1 = c_1 \qquad (10)$$
$$(t - p) r^2 = (T_0 - P_0) R_0^2 = c_2 \qquad (11)$$
$$(T_0 - P_0) R_0^2 = (T'_0 - P_1) R_1^2 = c_2 \qquad (12)$$

PART II.

THE SIMPLE ELASTIC STRENGTH OF A TUBE.

14. The *elastic strength* of a tube may be defined as the greatest resistance, expressed in tons or lbs. per square inch, which the tube can oppose within the elastic limit to the pressures acting over its surfaces.

In "SIMPLE ELASTIC STRENGTH", or the resistance of the tube to stress, we shall consider the separate effect of the stresses (tangential tensions, radial pressures or longitudinal tension) when acting independently and simultaneously.

A gun (tube) exerts its full elastic strength, as measured by *stress*, when both at the moment of firing and before firing, the maximum elastic stresses are equal to the limiting stresses permissible for the given design and material.

Later we shall discuss under the head of "COMPOUND ELASTIC STRENGTH" the effect of all of the stresses acting simultaneously to produce *strain*, or the resistance of the tube to *strain*, tangential, radial, or longitudinal, produced by the combined effect of the stresses.

The formulas to be deduced are derived from (9), (10), (11) and (12).

Combining (10) and (12) and eliminating T'_0, then substituting the resulting value of T_0 in (9) and (11) and eliminating p we find :

$$f = \frac{P_0 R_0^2 - P_1 R_1^2}{R_1^2 - R_0^2} + \frac{(P_0 - P_1) R_0^2 R_1^2}{R_1^2 - R_0^2} \cdot \frac{1}{r^2} \qquad - \qquad (13)$$

By a similar process we may deduce :

$$f = \frac{P_0 R_0^2 - P_1 R_1^2}{R_1^2 - R_0^2} - \frac{(P_0 - P_1) R_0^2 R_1^2}{R_1^2 - R_0^2} \cdot \frac{1}{r^2} \qquad - \qquad (14)$$

ELEMENTS OF ELASTIC STRENGTH OF GUNS.

If in the solution of (13) or any equation derived from it we find that t for any value of r is positive, then t is an extending tangential stress at the cylindrical surface considered; if t is negative the tangential stress is compressing, or $-t$ is a pressure. Therefore t may be either a tension or a pressure; p in (14) for all values of r is negative and therefore always a compressing stress or pressure.

Equation (13) gives the tension or tangential effect at any point of the tube corresponding to any assumed values of r, and (14) gives the pressure or radial effect at any point for any assumed value of r. When, in the transformation of these equations, it may be difficult to determine from the notation whether we are concerned with the tangential effect or the radial effect of the pressures P_0 and P_1, we will affect P_0 or P_1 with the suffix (θ) when we are considering a tangential effect, and with the suffix (ρ) when we are considering a radial effect. These suffices will be assumed when necessary and be omitted when unnecessary.

It should be remarked here that equation (13) is *the important equation* of this Part. The formulas of application in gun construction depend with very few exceptions upon (13) or equations derived from it; in other words, the elastic strength of a gun is determined almost wholly by formulas which give its resistance to *tangential rupture*.

By substituting for r in (13), R_0 and R_1 respectively, we will find the corresponding tangential tensions for the two surfaces of the tube in terms of the pressures. For the interior surface:

$$T_0 = P_0 \frac{R_1^2 + R_0^2}{R_1^2 - R_0^2} - P_1 \frac{2 R_1^2}{R_1^2 - R_0^2} \qquad (15)$$

and for the exterior surface:

$$T'_0 = P_0 \frac{2 R_0^2}{R_1^2 - R_0^2} - P_1 \frac{R_1^2 + R_0^2}{R_1^2 - R_0^2} \qquad (16)$$

Solving (15) with respect to P_0 we find:

$$P_0 = T_0 \frac{R_1^2 - R_0^2}{R_1^2 + R_0^2} + P_1 \frac{2 R_1^2}{R_1^2 + R_0^2} \qquad (17)$$

which formula has frequent application.

Although the notation of equations from (9) **to (17)** correspond to a separate tube or to the innermost tube of a built-up gun, yet they are derived from the general **formulas** (7) and (8) under conditions which would apply to any **tube**; therefore the notation of the preceding **formula,** or of any formulas derived from them, may be extended **to any** cylinder whatever its location in a built-up gun.

DISCUSSION OF "STRESS" EQUATIONS.*

15. In this investigation three cases may arise.

First.—When the exterior pressure is zero, or **may be so regarded** as compared with the interior pressure.

Second. When the interior pressure **is zero.**

Third. When both interior and exterior **pressures act.**

First Case: $P_1 = 0$.

16. **If the** exterior surface is **free** we may consider the pressure **on** it of the atmosphere **as zero,** $P_1 = 0$; then substituting this value in (13), **(14), (15), (16) and (17)** we find **(18), (19), (20) (21) and (22).**

$P_1 = 0$
$$t = \frac{P_0 R_0^2}{R_1^2 - R_0^2}\left(\frac{R_1^2 + r^2}{r^2}\right) \qquad (18)$$

$$p = -\frac{P_0 R_0^2}{R_1^2 - R_0^2}\left(\frac{R_1^2 - r^2}{r^2}\right) \qquad (19)$$

$$T_0 = P_0 \frac{R_1^2 + R_0^2}{R_1^2 - R_0^2} \qquad (20)$$

$$T'_0 = P_0 \frac{2 R_0^2}{R_1^2 - R_0^2} \qquad (21)$$

$$P_0 = T_0 \frac{R_1^2 - R_0^2}{R_1^2 + R_0^2} \qquad (22)$$

$$P_0 = \theta \frac{R_1^2 - R_0^2}{R_1^2 + R_0^2} \qquad (23)$$

$$T'_0 = T_0 \frac{2 R_0^2}{R_1^2 + R_0^2} \qquad (24)$$

$$R_1 - R_0 = R_0 \left(\sqrt{\frac{\theta + P_0}{\theta - P_0}} - 1\right) \qquad (25)$$

---

* The reader who wishes to investigate only the elastic strength of guns as determined by resistance to *strain*, may, with the exception of Article 21 and 22, omit from the beginning of Article 15 to Part III.

By comparing (18) and (19) we see that t for the same value of r will always be numerically greater than p, or that at any point of a tube subjected only to an interior pressure the tangential stress developed by this pressure will be greater than the radial pressure; hence in this case, since μ in gun steel is never less than θ, we need only investigate the tangential elastic resistance of the tube; t for all values of r is positive, hence represents an extending stress.

It is also apparent from (18) and (19) that t and p will each increase in value as r decreases and have their greatest value at the surface of the bore where $r = R_0$; hence the safety of the tube depends on the condition of its inner surface and is assured so long as T_0 does not exceed the elastic limit of the tube for extension; the least values of t and p are at the outer surface where $r = R_1$; in the latter case $p = 0 = P_1$.

From (20) we may obtain the tangential stress developed at the surface of the bore by a given interior pressure P_0.

No value of P_0 is admissible which gives a greater value to T_0 than θ.

Equation (21) gives the tangential stress developed at the outer surface of the tube by the pressure P_0.

From (22) we may obtain the pressure P_0 which will develop at the inner surface of the tube a tangential tension equal to any assigned value of T_0. The greatest value T_0 can have is θ and if we substitute it for T_0 in (22) we obtain (23).

This is the formula, (23), giving the elastic strength of a cast gun regarded as a simple homogeneous structure without initial tension.

We see also from equation (23)—though it is *not* a direct relation—that the elastic strength of a tube depends upon its relative dimensions; however, contrary to general belief, very little strength is gained by a considerable increase in the thickness of the walls of the tube. For example, let θ be the limit of the elastic strength of the metal employed, then from (23) we see that tubes which are 0.5, 1.0 and 1.5 calibres thick, will support interior pressures equal respectively to 0.6θ, 0.8θ and 0.88θ. These tubes being of equal length their weights are to each other

3, 8 and 15. That is while the tube which is 1.5 calibres thick weighs five times as much as the tube 0.5 calibres thick, yet there is only a gain in elastic resistance of less than 50 per cent. The pressure can only be increased to equal the tension when the tube is made infinitely thick for which $R_1 = \infty$, or $\frac{R_0}{R_1} = 0$.

Assuming $\theta = p$, equation (22) shows since T_0 must always be greater than P_0, that homogeneous guns burst not directly from the radial pressure P_0 of the powder gases, but from the tangential tensions developed by this pressure in the wall of the gun, and it also shows it is impossible to get from a homogeneous gun, however thick, a greater elastic resistance than the elastic strength of the metal employed.

As we increase the thickness of a tube the exterior elementary cylinders take but a very small share in the general work of resistance. For instance if we substitute for P_0 in (21) its value taken from (22) we will have (24); this expression gives the value or share of the resistance T'_0, at the exterior surface of the gun in terms of the resistance T_0 at the inner surface, when the gun is subjected to an interior pressure P_0. If the gun is half a calibre thick, $T'_0 = 0.4\ T_0$; if one calibre thick, $T'_0 = 0.2\ T_0$; if one and and-half calibres thick, $T'_0 = 0.12\ T_0$, and if infinitely thick, $T'_0 = 0\ T_0 = 0$.

Usually the thickness of wall in a cast gun is not greater than one and one-half calibres, and even with this thickness the interior elementary cylinder contributes over eight times as much as the exterior one to the general work of resistance.

In order to find the *thickness* necessary in a cast gun to resist a known interior pressure, solve (23) with respect to R_1, then subtract R_0 from each member and we will have (25).

17. The principal stresses are given by (18) and (19). These equations represent hyperbolas of the third degree which indicate at once the general form of the curves of tension and of pressure.

If we subject a tube having limiting radii R_0 and R_1 to an interior pressure P_0, we can by (18) construct a curve of tensions for all values of r.

In Figure 4 the ordinates DD_2, nn_2 and EE_1 represent the tensions corresponding to values of r for the respective abscissas CD, Cn and CE. The curve $D_2 n_2 E_1$ is the curve of tensions referred to the axes $CE - CC_3$.

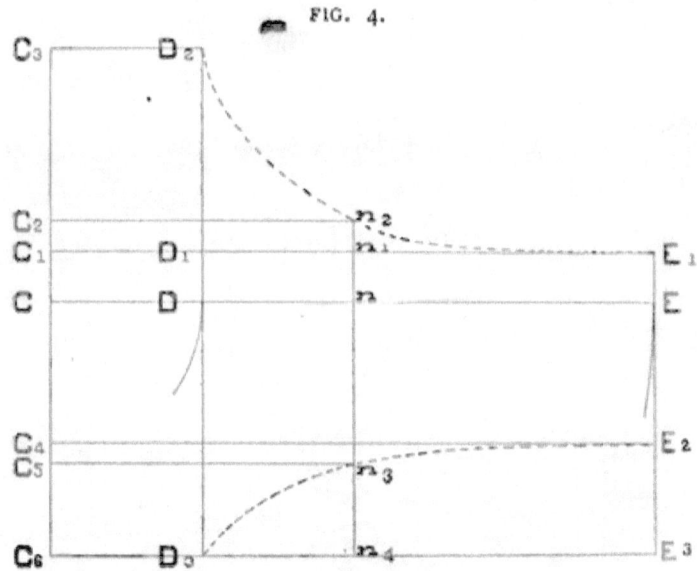

FIG. 4.

Since $p = T'_0 - t$, if through E_1 we draw $E_1 C_1$ parallel to CE, then the curve $D_2 n_2 E_1$ will represent the curve of pressures referred to the axes $C_1 E_1 - C_1 C_3$.

From a consideration of the figure we have:

$$CD = R_0 \qquad CE = R_1 \qquad Cn = r$$
$$DD_2 = T_0 \qquad EE_1 = T'_0 \qquad nn_2 = t$$
$$D_1 D_2 = P_0 \qquad (P_1 = 0) \qquad n_1 n_2 = p$$

The general condition of equilibrium of tubes of thickness $R_1 - R_0$, or $R_1 - r$ is:

$$P_0 R_0 = \int_{R_0}^{R_1} t\, dr; \qquad pr = \int_{r}^{R_1} t\, dr$$

In these expressions each first member is the rectangle of the radius of the inner surface considered, by the pressure at that surface.

The conditions of equilibrium are expressed geometrically by the equality of the following areas:

$$\text{Area } C_1 \, D_1 \, D_2 \, C_3 = \text{Area } D \, E \, E_1 \, D_2$$
$$\text{Area } C_1 \, n_1 \, n_2 \, C_2 = \text{Area } n \, E \, F_1 \, n_2$$

Second Case. $P_0 = 0$.

18. In the case when the tube is subjected to an exterior pressure only, we will have $P_0 = 0$. With this condition equations (13), (14), (15), (16) and (17) become (26), (27), (28), (29) and (30).

$P_0 = 0$
$$t = -\frac{P_1 R_1^2}{R_1^2 - R_0^2}\left(\frac{R_0^2 + r^2}{r^2}\right) \quad - \quad - \quad - \quad (26)$$

$$p = \frac{P_1 R_1^2}{R_1^2 - R_0^2}\left(\frac{R_0^2 - r^2}{r^2}\right) \quad - \quad - \quad - \quad (27)$$

$$(-T_0) = -P_1 \frac{2 R_1^2}{R_1^2 - R_0^2} \quad * \quad - \quad - \quad (28)$$

$$(-T'_0) = -P_1 \frac{R_1^2 + R_0^2}{R_1^2 - R_0^2} \quad * \quad - \quad - \quad (29)$$

$$P_1 = (-T_0)\frac{R_1^2 - R_0^2}{2 R_1^2} \quad - \quad - \quad - \quad (30)$$

$$P_1(\theta) = p \frac{R_1^2 - R_0^2}{2 R_1^2} \quad - \quad - \quad - \quad (31)$$

$$(-T'_0) = (-T_0)\frac{R_1^2 + R_0^2}{2 R_1^2} \quad - \quad - \quad - \quad (32)$$

$$R_1 - R_0 = R_0 \left(\sqrt{\frac{p}{p - 2 P_1}} - 1\right) \quad - \quad - \quad (33)$$

We see by comparing (26) and (27) that t is always negative and for the same value of r is always numerically greater than p; hence when a tube is subjected only to an exterior pressure we need only consider its tangential elastic strength.

We also see from (26) that t has its least numerical value when $r = R_1$ and increases numerically as r decreases until it has its greatest value when $r = R_0$.

---

* A tangential compressing stress at the inner surface of the tube will be represented by $(-T_0)$, and at the outer surface by $(-T'_0)$.

Negative tensions are pressures, and this means that all the elementary cylinders of the tube are in a state of tangential compression. Hence when a tube is subjected to an exterior pressure only, all the induced tangential stresses are compressing, and the stress of greatest numerical value is at the inner surface; therefore the safety of the tube is assured so long as the compressing stress $(-T_0)$ developed at the surface of the bore by P_1 does not exceed p in numerical value.

On the contrary it appears from (27) that p has its greatest numerical value at the outer surface and diminishes as r decreases until at the inner surface it is o.

Equation (28) gives the tangential compressing stress developed at the inner surface by an exterior pressure P_1. No value of P_1 is admissible which gives a greater value to $(-T_0)$ than p.

Equation (29) gives the tangential compressing stress developed at the outer surface of the tube by P_1.

From (30) we find the pressure P_1 which will develop at the inner surface of the tube a tangential compressing stress equal to any assigned values of $(-T_0)$.

The maximum value $(-T_0)$ can have is the elastic limit for compression, p; if we substitute p for $(-T_0)$ in (30) we obtain (31); this is the formula which gives the elastic strength of a tube to resist an external pressure.

Formula (31) is of especial importance since it limits the exterior pressure on the innermost cylinder or tube of a built-up gun in a state of rest; any greater exterior pressure in the state of rest would over compress the inner surface of the bore, and is therefore not admissible.

By combining (28) and (29) and eliminating P_1 we deduce (32) which gives the resistance of the elementary cylinder at the outer surface of the gun in terms of the resistance at the inner surface.

Assuming that the elastic limits for extension and for compression are equal, we see by comparing (23) and (31) that a tube will bear within the elastic limit a higher interior than exterior pressure.

In order to find the thickness necessary in a tube to resist a known exterior pressure, P_1 solve (31) with respect to R_1, then subtract R_0 from each member and we will have (33).

19. The curve of tangential compressions is given by (26) and the curve of pressures by (27).

In Figure (4) the curve $D_3 \, n_3 \, E_2$ represents the negative tensions referred to the axes $CE - CC_6$.

Since $p = T_0 - t$, if through D_3 we draw a line parallel to CE, then the curve $D_3 \, n_3 \, E_2$ will be the curve of pressures referred to the axes $C_6 \, E_3 - C_6 \, C$.   (FIG. 4: bis.)

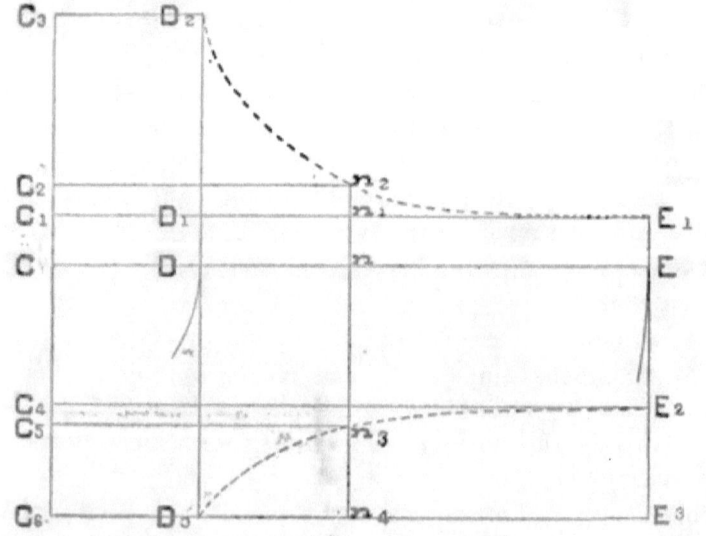

From a consideration of Figure 4 we have:

$C\,D = R_0$ $C\,E = R_1$ $C\,n = r$,
$D\,D_3 = (-T_0)$ $E\,E_2 = (-T'_0)$ $n\,n_3 = (-t_1)$
                                    $n_3 n_4 = p = (-T_0)-(-t)$
$(P_0 = 0)$ $E_2 E_3 = P_1$ $= DD_3 - n n_3$

$P_1 R_1 = -\int_{R_0}^{R_1} t\, dr = \text{Area } C_6 \, E_3 \, E_2 \, C_4 = \text{Area } D_3 \, E_2 \, E\, D$

$p\,r = -\int_{R_0}^{r} t\, dr = \text{Area } C_6 \, n_4 \, n_3 \, C_5 = \text{Area } D_3 \, n_3 \, n\, D$

It should be observed that the exterior pressures P_1 and p act respectively over the radii CE and Cn.

ELEMENTS OF ELASTIC STRENGTH OF GUNS.

Third Case: When P_0 and P_1 both act.

20. In this case the tube will be subjected both to an interior and to an exterior pressure.

We will assume equations (13), (14), (15), (16) and (17), and to preserve the sequence will renumber them.

$$t = \frac{P_0 R_0^2 - P_1 R_1^2}{R_1^2 - R_0^2} + \frac{(P_0 - P_1) R_0^2 R_1^2}{R_1^2 - R_0^2} \cdot \frac{1}{r^2} \quad (13) \quad (34)$$

$$p = \frac{P_0 R_0^2 - P_1 R_1^2}{R_1^2 - R_0^2} - \frac{(P_0 - P_1) R_0^2 R_1^2}{R_1^2 - R_0^2} \cdot \frac{1}{r^2} \quad (14) \quad (35)$$

$t > p \begin{cases} P_0 R_0^2 > P_1 R_1^2 \begin{cases} T_0 = P_0 \dfrac{R_1^2 + R_0^2}{R_1^2 - R_0^2} - P_1 \dfrac{2 R_1^2}{R_1^2 - R_0^2} & (15) \quad (36) \\[6pt] T'_0 = P_0 \dfrac{2 R_0^2}{R_1^2 - R_0^2} - P_1 \dfrac{R_1^2 + R_0^2}{R_1^2 - R_0^2} & (16) \quad (37) \\[6pt] P_0 = T_0 \dfrac{R_1^2 - R_0^2}{R_1^2 + R_0^2} + P_1 \dfrac{2 R_1^2}{R_1^2 + R_0^2} & (17) \quad (38) \\[6pt] P_0 = \theta \dfrac{R_1^2 - R_0^2}{R_1^2 + R_0^2} + P_1 \dfrac{2 R_1^2}{R_1^2 + R_0^2} & (39) \\[6pt] P_1(\theta) = P_0 \dfrac{R_1^2 + R_0^2}{2 R_1^2} - \theta \dfrac{R_1^2 - R_0^2}{2 R_1^2} & (40) \end{cases} \\[50pt] P_0 < P_1 \begin{cases} (-T_0) = -P_1 \dfrac{2 R_1^2}{R_1^2 - R_0^2} + P_0 \dfrac{R_1^2 + R_0^2}{R_1^2 - R_0^2} & (41) \\[6pt] (-T'_0) = -P_1 \dfrac{R_1^2 + R_0^2}{R_1^2 - R_0^2} + P_0 \dfrac{2 R_0^2}{R_1^2 - R_0^2} & (42) \\[6pt] P_1(\theta) = (-T_0) \dfrac{R_1^2 - R_0^2}{2 R_1^2} + P_0 \dfrac{R_1^2 + R_0^2}{2 R_1^2}* & (43) \\[6pt] P_1(\theta) = \rho \dfrac{R_1^2 - R_0^2}{2 R_1^2} + P_0 \dfrac{R_1^2 + R_0^2}{2 R_1^2} & (44) \end{cases} \end{cases}$

$t < p \begin{cases} P_0 > P_1 \\ P_0 R_0^2 < P_1 R_1^2 \end{cases} P_0(\rho) = \rho \qquad (45)$

$t = p \begin{cases} P_0 R_0^2 = P_1 R_1^2 \begin{cases} P_0(\theta) = \theta & (46) \\ P_0(\rho) = \rho & (47) \end{cases} \\ P_0 = P_1 \; \{ p = -t = \rho \qquad (48) \end{cases}$

---

* It should be remembered that $(-T_0)$ is a symbol to represent a tangential compressing stress at the surface of the bore and in applying (43) in its present form there should be substituted for $(-T_0)$ its positive numerical value.

In order to determine the location and value of **the greatest elastic stresses in a tube subjected to two pressures** P_0 **and** P_1 we can make the three suppositions.

$$t > p; \qquad t < p; \qquad t = p.$$

$$t > p.$$

1°. We see by comparing (34) and (35) that **the values of** t and p differ only by the sign of the expression

$$\frac{(P_0 - P_1) R_0^2 R_1^2}{R_1^2 - R_0^2} \cdot \frac{1}{r^2}$$

hence **if both terms of** t are **positive or both negative** $t > p$ numerically.

If $P_0 R_0^2 > P_1 R_1^2$ which involves **the condition that** $P_0 > P_1$ then will both terms of t **be positive and all** of **the** tube will **be** extended. If $P_0 < P_1$ which **involves the** condition $P_0 R_0^2 < P_1 R_1^2$, then will t be negative in both **terms and** all of the tube will be compressed. When $t > p$ **numerically we** are only concerned with the tangential **elastic strength of the tube.** An inspection of (34) shows that t whether an extending or **a** compressing stress has its greatest **numerical value at** the surface of the bore where $r = R_0$. Therefore **the condition of** the surface of the **bore is** the test of the safety **of the tube.** The values of t for **the inner and outer surfaces when** $t > p$ **and** all the tube **is extended are given by (36) and** (37) and **when all of** the tube **is compressed are given by (41)** and (42).

The greatest safe value of T_0 **is** θ, then if in (38) we substitute θ for T_0, we will obtain (39) **which** gives the tangential elastic strength **of a tube when** $t > p$ and **all the tube extended. If we wish a** greater margin of safety than working the tube to the full elastic limit **we** may by assigning reduced values to θ obtain reduced values **for** P_0.

For any **value** of P_0 in excess **of the** elastic strength of the simple tube as determined by **(23)**, equation (40) gives the value of the exterior **pressure,** P_1 which will, in connection with P_0, **cause** the surface **of the bore to be** extended exactly to its elastic limit; if any **less** value of P_1 be used, P_0 will cause the elastic strength of the tube to be exceeded.

ELEMENTS OF ELASTIC STRENGTH OF GUNS.

The greatest safe value ($=T_0$) can have is ρ; substituting ρ in (43) and solving for P_1 we get (44) which gives the tangential elastic strength of a tube when $t > p$ and all of the tube compressed, the inner surface being compressed to the elastic limit.

If we wish we can when all the tube is compressed obtain less values for P_1 by assigning less values to ρ.

$$t < p.$$

2°. The only reasonable hypothesis which will make $t < p$ is

and
$$\begin{matrix} P_0 > P_1 \\ P_0 R_0^2 < P_1 R_1^2 \end{matrix}$$

With this condition, both terms of p in (35) have the same sign while the terms of t in (34) have contrary signs, or the radial stress will at all points be greater than the tangential stress.

When $t < p$ we are only concerned with the radial stress to secure the safety of the tube. We see from (35) that with the preceding conditions, p increases numerically as r diminishes and has its greatest value at the surface of the bore. Hence when $t < p$, so long as the elastic limit of the tube is not exceeded in a radial direction at the surface of the bore, the safety of the tube is assured.

Since p is always negative, it is compressing in a radial direction and as the elastic limit of compression is ρ, if we substitute in (35), ρ for p and R_0 for r we get (45) which gives the radial elastic strength of a tube when $t < p$.

If one of the terms of t is positive and the other negative, then t may change its sign as we diminish r from R_1 to R_0, and t will be equal to zero when the following condition is satisfied:

$$(P_0 R_0^2 - P_1 R_1^2) r^2 + (P_0 - P_1) R_0^2 R_1^2 = 0$$

or
$$r = R_0 R_1 \sqrt{\frac{P_0 - P_1}{P_1 R_1^2 - P_0 R_0^2}} \qquad (49)$$

The elementary cylinder with a radius equal to this deduced value of r will be neutral with respect to tangential tension; that is without tangential extension or compression; since t increases algebraically as we diminish r, all of the tubes within the neutral cylinder will be extended and all beyond will be compressed.

$$t = p.$$

3°. t can only be equal numerically to p under the supposition that:
$$P_0\, R_0^2 = P_1\, R_1^2$$
or
$$P_0 = P_1$$

With the first condition we see that the greatest value of t and p is at the inner surface and to obtain the elastic strength of the tube substitute in (34) and (35) R_0 for r, and $\dfrac{P_0\, R_0^2}{R_1}$ for P_1, and we shall obtain:

$$P_0 = T_0 : \text{ and } P_0 = P_0 :$$

replacing the second members by the elastic limits, we have (46) and (47). If the two elastic limits are not equal, we take the less value as the elastic strength of the tube.

With the second condition t and p are at all points numerically equal and we will obtain the elastic strength of the tube from (48).

A review of this discussion shows that the greatest stress in numerical value is always at the interior surface of the tube, and therefore the condition of the innermost elementary cylinder is really the criterion of the resistance of the tube; if its elastic strength be not exceeded the integrity of the tube is assured.

TENSIONS AND PRESSURES AT REST, AND OF FIRING.
NATURAL TENSIONS, AND PRESSURES OF FIRING.

21. It will be convenient to develop here the relation existing between tensions (or pressures) of firing, at rest and the natural tensions (or pressures) of firing.

It is evident from the fundamental principle enunciated in Article 10 that if a homogeneous tube be subjected simultaneously to two pressures, one exterior and the other interior, the complete effect of the two pressures, or tensions at any point will be obtained by taking the algebraic sum of the components.

According to the principle just stated, the tension (or pressure) which is developed at any point of a built-up gun (compound tube) at the moment of firing, is the algebraic sum of that which exists in a state of rest and that which the powder gases would produce in the gun if no initial tension existed in a

ELEMENTS OF ELASTIC STRENGTH OF GUNS. 31

state of rest; the latter condition of "no initial tension" makes the gun a homogeneous tube.

The following general property may be enunciated. *The tension* (or *the pressure*) *at any point of a built-up gun in the state of firing is the algebraic sum of the tension* (or *of the pressure*), *which exists at this point in a state of rest, and of that which would be produced under the action of the same interior pressure in a homogeneous gun whose dimensions are equal to the total dimensions of the tube and hoops of the built-up gun.*

This tension (or pressure) in a homogeneous gun of equal dimensions may be called the *natural tension* (or *pressure*) *of firing*. Hence we deduce the simple rules.

The tension of firing is the algebraic sum of the natural tension of firing and the tension at rest.

The tension at rest is the algebraic difference of the tension of firing and the natural tension of firing.

The natural tension of firing is the algebraic difference of the tension of firing and the tension at rest.

Similar rules apply to pressures.

It is of course evident that the pressure of firing and the natural pressure of firing are the same in the bores of the built-up gun and of the homogeneous gun.

As we shall have occasion to use this principle with respect to pressures in deducing shrinkage formulas for the state of rest, we will express it algebraically.

Let us represent by:

$P_0, P_1, P_2, \ldots P_m \ldots P_n$, pressures of firing.
$P'_0, P'_1, P'_2 \ldots P'_m \ldots P'_n$, pressures at rest.
$p_0, p_1, p_2 \ldots p_m \ldots p_n$, natural pressures of firing.
$P'_0 = 0; \quad P_n = P'_n = p_n = 0.$
$T_0, T_1, T_3, \ldots T_m, \ldots T_n$, natural pressure of firing.

$$\left. \begin{array}{lr} P_0 = P_0, & a \\ P_1 = P'_1 + p_1, & b \\ \cdots\cdots\cdots\cdots\cdots & \\ P_m = P'_m + p_m, & m \\ \cdots\cdots\cdots\cdots\cdots & \\ P_{n-1} = P'_{n-1} + p_{n-1}, & n-1 \end{array} \right\} \quad (50)$$

$$\left.\begin{array}{ll} P'_0 = 0, & a \\ P'_1 = P_1 - p_1, & b \\ \cdots\cdots\cdots\cdots & \\ \cdots\cdots\cdots\cdots & \\ P'_{n-1} = P_{n-1} - p_{n-1}, & n-1 \end{array}\right\} \quad (51)$$

$$\left.\begin{array}{ll} P_0 = P_0, & a \\ p_1 = P_1 - P'_1, & b \\ \cdots\cdots\cdots\cdots & \\ \cdots\cdots\cdots\cdots & \\ p_{n-1} = P_{n-1} - P'_{n-1} & n-1 \end{array}\right\} \quad (52)$$

Similar equations might be written for tensions.

A COMPOUND TUBE, AND INITIAL TENSIONS.

22. We may construct a compound tube of two concentric cylinders, in which the outer cylinder is superposed by *shrinkage*, and the compound tube so assembled will have a greater elastic strength to resist interior pressures than a simple tube of the same dimensions.

This operation of shrinkage consists in making the outer diameter of the inner cylinder slightly greater than the inner diameter of the outer cylinder; then dilating the outer cylinder—generally by heat—until it may be made to enclose the inner cylinder. The inner cylinder or tube is said to be *hooped* and the outer cylinder is called the *hoop*. The long hoop over the tube of a gun is usually called the jacket. A compound tube is thus obtained which in a state of rest—that is without pressure on exterior, or interior surface—presents the following properties:

The outer cylinder subjected to an interior pressure will be dilated or in a state of initial tangential extension; the inner cylinder subjected to an exterior pressure will be contracted or in a state of initial tangential compression. It is of course evident that at the common surface of contact, the pressure P'_1 due to shrinkage acts radially with equal intensity both inward and outward.

After assemblage the two cylinders will have a common surface of contact with radius R_1.

The tangential stresses of extension and compression existing in a compound tube in a state of rest are usually called *initial tensions*.

FIG. 5.

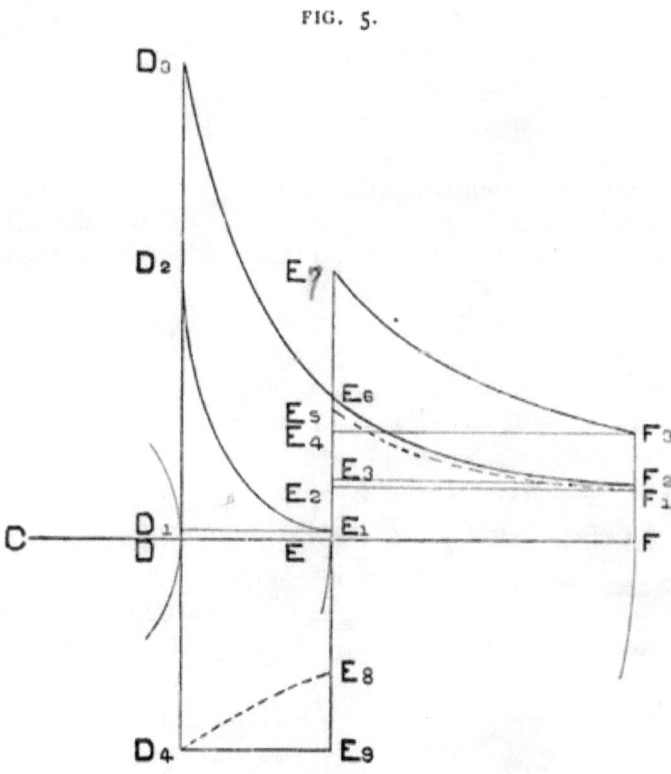

The curves of initial tensions developed in the two cylinders will be represented by $E_5 F_1$ and $D_4 E_8$ (Figure 5). Initial tensions, will always be represented by a broken line. The curve of tensions for the hoop can be constructed from (18) and the curve of compressions (negative tensions) for the inner tube from (26).

In the compound tube we have:

$$CD = R_0 \qquad CE = R_1 \qquad CF = R_2$$
$$DD_4 = (-T_0) \qquad EE_8 = (-T'_0) \qquad FF_1 = T_1$$
$$(P_0 = 0) \qquad EE_5 = T_1 \qquad (P_2 = 0)$$
$$E_8 E_9 = E_2 E_3 = P_1$$

Elements of elastic strength of guns. 5.

Since the two cylinders which form the compound tube are in equilibrium, the resistance of the compressions developed in the inner tube must be equal to the resistance of the tensions developed in the hoop, or:

$$P_1 \; R_1 = \text{Area } DD_4 E_8 E = \text{Area } E F F_1 E_5 *$$

Let us suppose the interior of the compound tube subjected to a pressure P_0 sufficiently great to dilate the interior surface of the inner tube from a state of initial compression $(-T_0)$ to a state of tangential tension equal to T_0, or a range of tangential stress which in a homogeneous tube without initial tension would be equal to a tangential tension represented by the arithmetical sum of $(-T_0)+T_0$. †

The curve of natural tensions of firing is found as follows:

Substitute in (22) for T_0 the arithmetical sum of $(-T_0)$ and T_0 and for R_1, the exterior radius R_2 and thus obtain the natural pressure P_0. With this value of P_0 in (18), notation extended by substituting R_2 for R_1, find the value of t corresponding to R_1, and thus obtain $\overline{T_1}$. We now have the natural tensions

$$\overline{T_0} = T_0 - (-T_0); \quad \overline{T_1} \text{ from (18)}; \quad \overline{T_2} = T_0 - P_0.$$

The curve of natural tensions will be $D_3 \; E_6 \; F_2$.

The curve of tensions of firing of the compound tube is found from the natural tensions, and tensions at rest as follows (Article 21):

$$T_0 = DD_3 - DD_4 = DD_2 \qquad T'_0 = EE_6 - EE_8 = EE_1$$
$$T_1 = EE_6 + EE_5 = EE_7 \qquad T'_1 = FF_2 + FF_1 = FF_3$$

The curve of tensions of firing will be $D_2 \; E_1$ and $E_7 \; F_3$.

We may conceive then that by the *judicious selection* of a shrinkage we can cause the interior surfaces of the two cylinders of the compound tube subjected to an interior pressure P_0, to receive simultaneously the maximum value of the pressures or tensions to which they can be subjected without permanent deformation.

By these means we should obtain the maximum elastic resistance of the compound tube for the radii R_0, R_1 and R_2.

---

* If we determine these areas by the integral calculus we will find them equal with contrary signs.

† $(-T_0)$ and T_0 may, or may not be numerically equal. $(-T_0)$ is simply a symbol to represent a stress of compression.

We have in Figure 5 the following conditions:
$$P_0 R_0 = \text{Area } DEE_1 D_2 + \text{Area } EFF_3 E_7$$
$$= \text{Area } D_4 E_8 E_1 D_2 + \text{Area } E_5 F_1 F_3 E_7.$$

Hence the total resistance to the pressure P_0 is equal to the sum of the areas of extension in the two tubes.

We see since $E_4 E_7$ is equal to $E_2 E_5 + E_3 E_6$ (Article 21) that the pressure at the contact surface is greater when P_0 acts than when the system is at rest.

The pressure P'_1 exerted by the inner tube upon the hoop in the state of rest is equal to the reaction of the hoop, acting as an exterior pressure upon the inner tube, but the tensions developed at the two surfaces of contact are very different:

$$EE_1 = T'_0 ; \quad EE_7 = T_1.$$

Thus we see that in the compound tube the *development* of the elastic stresses *is quite different* from that which would be produced in a homogeneous tube. But in accord with the principle assumed in Article 10 of the independence of the effect of forces acting simultaneously, the *variations* of the elastic pressures and tensions with the interior pressure P_0 *are absolutely the same* in the compound tube and in the homogeneous tube, for the same value of r, provided the dimensions are identical and the moduli of elasticity the same. In fact the object of Article 21 was to illustrate this principle and show its practical application.

THE ELASTIC STRENGTH OF A TUBE WITH INITIAL TENSIONS. THE NATURE OF REINFORCEMENT.

23. The object of assembling a compound tube by shrinkage is to get a greater resistance to an interior pressure than can be obtained from a simple tube of the same dimensions.

In a built-up gun consisting of a tube and other cylinders there is developed, in a state of rest, due to the contractile force of the jacket and hoops, a tangential stress of compression at the inner surface of the tube. On firing, the pressure of the powder gases extends all of the elementary cylinders of the simple tube from their compressed condition in the state of rest. It is evident that a compressed tube can be extended through a greater range without exceeding the elastic limit for extension than a neutral tube, that is, one without initial tension.

The nature of reinforcement of a tube depends on the principle that if a tube is *compressed* at its inner surface to its elastic limit, or to a less limit, it can, with safety, be *extended* to its elastic limit, that is, through a range of $p+\theta$, or through the less range of $(-T_0)+\theta$.

Equation (39) gives the elastic strength of a compound tube subjected at the instant of firing to an exterior pressure P_1 on the inner cylinder (tube), and in which there is developed at the inner surface of the inner tube by the action of P_0 and P_1, the limiting tangential stress of extension θ. In other words (39) expresses the condition that in the state of firing, the innermost cylinder of the built-up gun shall be tangentially extended to its elastic limit.

If we examine the two terms of P_0 in (39) we see the first is a maximum since it gives the elastic strength of a simple tube with radii, R_0 and R_1, (23); the second term is the expression for the compressing stress $(-T_0)$ developed in a tube with the same radii subjected to the exterior pressure P_1, (28).

P_0 will then be a maximum when P_1 is a maximum, but the firing pressure P_1 is frequently limited by the maximum admissible pressure P'_1 which exists before firing at the first shrinkage surface. The maximum value of P'_1 is limited by (31) since any greater pressure at rest would over compress the surface of the bore. The value of the pressure of firing P_1 is equal to the pressure at rest, plus the natural pressure of firing (50). It is the firing pressure P_1 which in (39) determines the elastic strength of a built-up gun. Let us assume that P'_1 at rest compresses the tube to its elastic limit, then when P_0 and P_1 both act, the tube is extended at the surface of the bore from the limit of compression to the limit of extension.

The value of P'_1 found from (31) is the limiting exterior pressure for the state of rest; if we substitute this value of P'_1 for P_1 in (39) we obtain after reduction:

$$P_0 = (p+\theta)\frac{R_1^2 - R_0^2}{R_1^2 + R_0^2} \tag{53}$$

This expression is independent of the exterior pressure. As applied to gun construction, it gives the elastic strength of a simple tube, (23), with thickness of wall equal to $R_1 - R_0$, in which before firing there is, at the surface of the bore, a tangential compressing stress equal to p and in which in the state of firing there is developed the tangential tension θ; or in other words this formula shows if a simple tube in a state of rest be *compressed* tangentially to its elastic limit at the surface of the bore, that the interior pressure P_0, which develops in this tube at its inner surface a tangential tension equal to the elastic limit, would produce in a simple tube of the same dimensions without initial tension, a tangential extending stress equal to $p+\theta$.

If $p=0$, we see that the elastic strength of the tube given by (53) is twice as great as the elastic strength given by (23).

A gun may be said to be "perfectly reinforced" when, before firing, $(-T_0)=p$, and at the instant of firing, $T_0=0$.

If $R_1 = \infty$, then will the gun be infinitely thick and:

$$P_0 = p + \theta \qquad (54)$$

which is the limit of the strength of a "perfectly reinforced" gun.

Equation (53) gives the elastic strength of a "naturally hooped" gun, where the initial compression is secured as in the Rodman process by casting hollow and cooling from the interior, or where the initial tension is obtained as in the Uchatius phosphor-bronze guns, by mandrelling.

It should be stated that the production of initial tension by the methods indicated has not been wholly satisfactory, or in other words it would not be proper to substitute the value of p, as determined by physical tests, in (53) in order to find the elastic strength of such a gun.

The pressure of firing, P_1 in a "naturally hooped" gun is o.

In a built-up gun the compressing stress at the surface of the bore in the state of rest is caused by the pressure P'_1, but P'_1 itself results from the contractile force of the cylinders beyond R_1.

Equation (39) gives the elastic resistance of a built-up gun with thickness of wall *greater* than $R_1 - R_0$; the firing pressure P_1 is equal to the sum of the pressure due to the contractile force of the cylinders assembled beyond R_1 and to the natural pressure of firing at R_1.

If a built-up gun has been properly assembled by shrinkage it may thereafter be treated as a simple homogeneous tube; the theory of reinforcement by shrinkage assumes a compressing stress at the surface of the bore, which may or may not reach the elastic limit; and the theory further gives to the gun an elastic strength which is the powder pressure which will extend the surface of the bore from the limit of its tangential compressing stress to the limit of the stress for tangential extension; this principle may be expressed algebraically by extending the notation of (23) to include the limiting radii of the gun and by substituting $(-T_0)^* + \theta_0$ for θ; or if the compressing stress is equal to the elastic limit by substituting $\rho_0 + \theta_0$ for θ.

If the exterior radius is R_2 we will have from (23) for the elastic strength of a perfectly reinforced gun composed of tube and jacket:

$$P_0 = (\rho_0 + \theta_0) \frac{R_2^2 - R_0^2}{R_2^2 + R_0^2} \tag{55}$$

If the exterior radius is R_n the formula becomes by extending the notation of (23):

$$P_0 = (\theta_0 + \rho_n) \frac{R_n^2 - R_0^2}{R_n^2 + R_0^2} \tag{56}$$

Equation (56) gives the elastic strength of a built-up gun with the condition that *before firing* the surface of the bore is compressed to the elastic limit.

The value of P_1 in (39) is dependent on P'_1 and if P'_1 compresses the surface of the bore exactly to the elastic limit, then P_0, the elastic strength of the gun, will have the same value in (39) and (56).

If the compressing stress at the surface of the bore, in the state of rest, is less than ρ, the elastic strength of the gun will be given by:

$$P_0 = [(-T_0)^* + \theta_0] \frac{R_n^2 - R_0^2}{R_n^2 + R_0^2} \tag{57}$$

When P_1 is a maximum the gun may be assembled so that, in the state of firing, each simple cylinder shall be extended to its elastic limit.

---

* The positive numerical value of $(-T_0)$ **must** be substituted in the formula.

The elastic strength of a built-up gun may, then, be given by two equations; one of which, (39), expresses the condition that in the firing state each cylinder shall exert its full elastic strength for extension; the other, (56), expresses the condition that in the firing state the surface of the bore shall be extended to its elastic limit, but contains the condition that in the state of rest the surface of the bore shall be compressed to the elastic limit. Later, in Article 29, we will explain which of these is the firing pressure to be selected for the deduction of the shrinkages to be applied in the construction of the gun.

This selected value of P_0 will be the maximum service pressure which might be used in the gun.

THE ELASTIC RESISTANCE OF A BUILT-UP GUN IN TERMS OF THE TANGENTIAL TENSIONS.

24. Let us suppose the gun assembled by shrinkage, and to contain n simple cylinders.

Let us represent by:

$R_0, R_1, \ldots R_n$, limiting radii of the simple cylinders.

$P_0, P_1, \ldots P_n$, pressures at the instant of firing corresponding to these radii.

$T_0, T_1, \ldots T_{n-1}$ tensions of firing developed at the inner surfaces of the cylinders.

P_0 is the firing pressure in the bore.

$P_1, P_2 \ldots P_{n-1}$ are the firing pressures at the shrinkage surfaces, due to the combined action of P_0 and the contractile action of the envelope beyond the particular surface considered.

$P_n = 0$, as it is the pressure of the atmosphere.

Equation (38) gives the elastic resistance of the inner tube in action:

$$P_0 = T_0 \frac{R_1^2 - R_0^2}{R_1^2 + R_0^2} + P_1 \frac{2 R_1^2}{R_1^2 + R_0^2}$$

P_1 acts with equal intensity inward and outward, and considered as acting outward, it gives the elastic resistance of the second cylinder when subjected to the two pressures P_1 and P_2. We may find its value in terms of the second cylinder by extending the notation of (38) and obtain:

$$P_1 = T_1 \frac{R_2^2 - R_1^2}{R_2^2 + R_1^2} + P_2 \frac{2R_2^2}{R_2^2 + R_1^2}$$

Let us assume $n=3$, whence $P_3=0$.

Finding the value of P_2 in terms of the third cylinder, we have:

$$P_2 = T_2 \frac{R_3^2 - R_2^2}{R_3^2 + R_2^2}$$

Eliminating from these equations P_2 and P_1, we get:

$$\begin{aligned} P_0 &= T_0 \frac{R_1^2 - R_0^2}{R_1^2 + R_0^2} + T_1 \frac{2R_1^2}{R_1^2 + R_0^2} \cdot \frac{R_2^2 - R_1^2}{R_2^2 + R_1^2} \\ &+ T_2 \frac{2R_1^2}{R_1^2 + R_0^2} \cdot \frac{2R_2^2}{R_2^2 + R_1^2} \cdot \frac{R_3^2 - R_2^2}{R_3^2 + R_2^2} \end{aligned} \quad (58)$$

The law of formation of the separate terms is evident. If the gun consists of n cylinders, the formula will be:

$$\left.\begin{aligned} P_0 &= T_0 \frac{R_1^2 - R_0^2}{R_1^2 + R_0^2} + T_1 \frac{2R_1^2}{R_1^2 + R_0^2} \cdot \frac{R_2^2 - R_1^2}{R_2^2 + R_1^2} \\ &+ T_2 \frac{2R_1^2}{R_1^2 + R_0^2} \cdot \frac{2R_2^2}{R_2^2 + R_1^2} \cdot \frac{R_3^2 - R_2^2}{R_3^2 + R_2^2} \\ &+ T_3 \frac{2R_1^2}{R_1^2 + R_0^2} \cdot \frac{2R_2^2}{R_2^2 + R_1^2} \cdot \frac{2R_3^2}{R_3^2 + R_2^2} \cdot \frac{R_4^2 - R_3^2}{R_4^2 + R_3^2} \\ &+ \ldots \ldots \ldots \ldots \ldots \ldots \\ &+ \ldots \ldots \ldots \ldots \ldots \ldots \\ &+ T_{n-1} \frac{2R_1^2}{R_1^2 + R_0^2} \cdot \frac{2R_2^2}{R_2^2 + R_1^2} \cdots \frac{2R_{n-1}^2}{R_{n-1}^2 + R_{n-2}^2} \cdot \frac{R_n^2 - R_{n-1}^2}{R_n^2 + R_{n-1}^2} \end{aligned}\right\} \quad (59)$$

which gives the resistance of the gun in terms of the positive tensions T_0, T_1, T_2 ... and T_{n-1} at the inner surfaces of the separate cylinders. When T_0, T_1 ... T_2 and T_{n-1} are equal to the elastic limit, the value of P_0 will be a maximum.

In the expression for P_0 in (59), the first term is the resistance of the tube without initial compression, the second is the share of the second cylinder, and so on, in the general resistance of the wall to the pressure P_0.

It is evident that we may assign various values to T_0, T_1 ... T_{n-1} not in excess of the elastic limit of the cylinder considered, and thus obtain admissible values of P_0.

Or, as will be shown in Article 27,—if the gun is in excess of a definite thickness for the number of cylinders employed—we may obtain the same admissible maximum value for P_0 by assigning values to all the tensions except one, and solving, find the value of that special one.

These values of P_0, for the state of firing are, however, limited by the condition that the pressures in the state of rest due to the shrinkage, shall not over compress the inner surface of any cylinder.

THE MOST ADVANTAGEOUS RADIUS FOR SHRINKAGE SURFACE.

25. Having a gun of definite calibre, thickness and number of cylinders, it is now wished to determine the intermediate radii which will give the maximum resistance.

Let us assume any two adjacent cylinders and denote by :

R, interior radius of the inner cylinder.
r, radius of the surface of contact of the two cylinders.
R', exterior radius of the outer cylinder.
P, firing pressure corresponding to radius R.
p, firing pressure corresponding to radius r.
P', firing pressure corresponding to radius R'.
θ and θ', elastic limits of the cylinders.

We will assume R and R' constant while r varies.

If the resistance of the two cylinders at the instant of firing is a maximum, the tangential stresses developed will be θ and θ'.

Applying (36) and changing the notation to correspond to this case, we have:

$$\theta = P \frac{r^2 + R^2}{r^2 - R^2} - p \frac{2r^2}{r^2 - R^2}$$

$$\theta' = p \frac{R'^2 + r^2}{R'^2 - r^2} - P' \frac{2 R'^2}{R'^2 - r^2}$$

Eliminating p from these equations and solving with respect to P, we obtain:

$$P = \theta \frac{r^2 - R^2}{r^2 + R^2} + \theta' \frac{2r^2}{r^2 + R^2} \cdot \frac{R'^2 - r^2}{R'^2 + r^2} + P' \frac{2r^2}{r^2 + R^2} \cdot \frac{2 R'^2}{R'^2 + r^2}$$

P' is constant since it depends only on the elastic limits and dimensions of the cylinders outside of the two cylinders considered, and by hypothesis these cylinders are unchanged.

R and R' are constant, therefore P depends on r.

The value of P_0 depends on P, and on it exclusively, since the cylinders within the two considered remain unchanged.

Therefore the gun will have a maximum resistance when P is a maximum.

Reducing the expression for P to a common denominator, we obtain:

$$P = \frac{r^4(\theta - 2\theta') + r^2\left[(R'^2 - r^2)\theta + 2R'^2\theta' + 4R'^2 P'\right] - R'^2 R^2 \theta}{r^4 + r^2(R'^2 + R^2) + R'^2 R^2}$$

Differentiating and placing the first differential coefficient with respect to r^2 equal to 0, we obtain:

$$r^4\left[2R'^2 P' + \theta'(2R'^2 + R^2) - \theta R^2\right]$$
$$- 2r^2 R'^2 R^2 (\theta - \theta') - R'^4 R^2 (2P' + \theta + \theta') = 0.$$

Placing

$$2R'^2 P' + \theta'(2R'^2 + R^2) - \theta R^2 = A$$
$$R'^2 R^2 (\theta - \theta') = B$$
$$R'^4 R^2 (2P' + \theta + \theta') = C$$

and solving for r^2, we get:

$$r^2 = \frac{B \pm \sqrt{B^2 + A^2 C}}{A} \tag{60}$$

C is always positive, therefore $A^2 C$ will be positive and the sign of r^2 will depend on the sign of the radical, but as r^2 cannot be negative, the $+$ sign of the radical must be taken.

Hence if there is a value of r giving a maximum resistance of the gun its square is given by the positive root of (60).

If the elastic limit in both cylinders is the same, $\theta = \theta'$, and we deduce:

$$r^2 = R R' \tag{61}$$

Hence if two adjacent cylinders in a built-up gun have the same elastic limit, the most advantageous intermediate radius will be a mean proportional between the extreme radii.

In practice other considérations cause a frequent modification of this rule. The cylinder carrying the breech-block and which therefore sustains the longitudinal stress is generally made of a greater thickness than would follow from the law just enunciated. Also if we take the most advantageous radii, the thickness of the simple cylinders might be so small that it would be difficult to prepare them.

TANGENTIAL ELASTIC STRENGTH OF A BUILT-UP GUN WITH MOST ADVANTAGEOUS RADII AND CONSTANT ELASTIC LIMITS.

26. When all the cylinders of a built-up gun have a common elastic limit and the radii of the cylinders for a given thickness of wall are the "most advantageous" we can reduce (59) to a much simpler form.

When the elastic limits are constant in order that the resistance of the gun for assumed values of R_0 and R_n shall be a maximum, we must have:

$$\left.\begin{array}{c} R_1^2 = R_2 R_0 \\ R_2^2 = R_3 R_1 \\ R_3^2 = R_4 R_2 \\ \cdots \cdots \cdots \\ \cdots \cdots \cdots \\ R_{n-1}^2 = R_n R_{n-2} \end{array}\right\} \quad (62)$$

Let us represent the ratio $\dfrac{R_0}{R_1}$ by a; then:

$$R_1 = \frac{1}{a} R_0$$

Substituting this value of R_1 in the first equation of (62) we get:

$$R_2 = \left(\frac{1}{a}\right)^2 R_0$$

Substituting this value in the second equation of (62) we get:

$$R_3 = \left(\frac{1}{a}\right)^3 R_0$$

Proceeding in this manner with all the equations in (62) we obtain:

$$\left.\begin{array}{l}R_1=\left(\dfrac{1}{a}\right)R_0 \\ R_2=\left(\dfrac{1}{a}\right)^2 R_0 \\ R_3=\left(\dfrac{1}{a}\right)^3 R_0 \\ \cdots\cdots\cdots \\ \cdots\cdots\cdots \\ R_{n-1}=\left(\dfrac{1}{a}\right)^{n-1} R_0 \\ R=\left(\dfrac{1}{a}\right)^{n} R_0\end{array}\right\} \qquad (63)$$

Thus we see that the most advantageous values of the radii of the cylinders intermediate between R_0 and R_n form (the elastic limit being constant for all the cylinders) a geometrical progression.

The denominator a of the ratio in this progression is determined from the last equation of (63), and is:

$$a=\left(\dfrac{R_0}{R_n}\right)^{\frac{1}{n}} \qquad (64)$$

Therefore the values of $R_1, R_2, \ldots R_{n-1}$ dependent on R_0 and R_n will be:

$$R_1=\left(\dfrac{R_n}{R_0}\right)^{\frac{1}{n}} R_0 \ ; \quad R_2=\left(\dfrac{R_n}{R_0}\right)^{\frac{2}{n}} R_0 \ ; \quad \cdots \quad R_{n-1}=\left(\dfrac{R_n}{R_0}\right)^{\frac{n-1}{n}} R_0.$$

Substitute in (59) for $R_1, R_2, \ldots R_{n-1}$ their values taken from (63) and for $T_0, T_1 \ldots T_{-1}$ their common value θ and we will have:

$$P_0=\theta\dfrac{1-a^2}{1+a^2}\left[1+\left(\dfrac{2}{1+a^2}\right)+\left(\dfrac{2}{1+a^2}\right)^2+\cdots+\left(\dfrac{2}{1+a^2}\right)^{n-1}\right]$$

Within the brackets we have a geometrical progression in which the ratio is $\dfrac{2}{1+a^2}$. Taking the sum of the series we get:

$$P_0=\theta\left[\left(\dfrac{2}{1+a^2}\right)^n - 1\right] \qquad (65)$$

ELEMENTS OF ELASTIC STRENGTH OF GUNS. 45

THE MOST SUITABLE THICKNESS OF WALL OF A PERFECTLY REINFORCED GUN, WITH CONSTANT ELASTIC LIMITS.

27. If we suppose the gun composed of n cylinders, all having the same elastic limit and conforming to the law of the most advantageous radii, then the elastic strength of the gun is determined from (65) which is deduced under the hypothesis that at the moment of firing, all the cylinders are worked to their elastic limit:

Placing
$$v = \left(\frac{R_0}{R_n}\right)^{-2} = a^2 \tag{66}$$

we deduce from (65):
$$P_0 = \theta \left\{ \frac{2^n}{(1+v)^n} - 1 \right\} \tag{67}$$

In this, v fixes the thickness of the wall in calibres.

The elastic strength of the gun determined from the condition that at rest the surface of the bore is compressed to the elastic limit, and in firing the same surface is extended to the elastic limit, is given by (56):
$$P_0 = (p+\theta)\frac{R_n^2 - R_0^2}{R_n^2 + R_0^2}$$

Placing $p = 0$, and observing that
$$R_n^2 = \frac{R_0^2}{v^n} \tag{68}$$

we get:
$$P_0 = 2\theta \frac{1-v^n}{1+v^n} \tag{69}$$

Equating (67) and (69); we find:
$$(1+v)^n (3-v^n) = 2^n (1+v^n) \tag{70}$$

From (70), having assumed the number n of cylinders, we can determine v, which enables us to find the exterior radius R from (68). This value of R_n with the known value of R_0 substituted in (67) and (69) will satisfy both. Therefore this value of R satisfies, first, the condition that in the state of firing, all the cylinders are extended to their elastic limit and thus perform their full share of the work, second, that in the state of rest the surface of the bore is compressed to the elastic limit.

If R_n is greater than the value deduced from (68) and we compel all the cylinders in firing to exert their full elastic strength, then in a state of rest, the surface of the bore will be compressed beyond the elastic limit.

On the other hand, if at rest the surface of the bore is compressed to the elastic limit then in the state of firing, all the cylinders will not be extended to their elastic limit, or the full elastic strength of all the cylinders will not be exerted.

If R_n is less than the value deduced from (68) and it is wished to have all the cylinders exert their full elastic strength in the state of firing, then before firing, the surface of the bore will not be compressed to its elastic limit ; on the other hand, if the surface of the bore in a state of rest is compressed to its elastic limit, then at the instant of firing, if the tube be worked to its elastic limit, the other cylinders will be extended beyond their elastic limit.

R_n deduced from (68) determines the thickness of wall, which is necesssary in order that in a state of rest the surface of the bore shall be compressed to its elastic limit and in the state of firing all of the cylinders shall be worked to their elastic limit if the gun be properly assembled.

Of the n roots of (70) we see by inspection that one is always equal to unity, but it is apparent from (66) that this value cannot be used since R_n is never equal to R_0.

Since R_n is greater than R_0, no value of v can be used except a real positive value less than unity. Assuming $n=2$, we find that:
$$v = 0.18$$
whence :
$$R_n = \frac{1}{0.18} R_0 = 5.56 R_0$$
consequently the thickness of the wall in calibres will be :
$$\frac{R_n - R_0}{2 R_0} = 2.28 \qquad (71)$$

This being the case, the elastic strength of the gun calculated by formula (67) or (69) will be:

$$P_0 = 1.87\, \theta. \qquad (72)$$

On the supposition that $n=3$, we find:

$$v^{\frac{3}{2}} = 0.28 \quad \therefore\ v = 0.43$$

and
$$R_u = \frac{1}{0.28} R_0 = 3.57\, R_0$$

consequently the thickness of the wall in calibres will be:

$$\frac{R_u - R_0}{2R_0} = 1.28 \qquad (73)$$

This being the case the elastic strength of the gun by formula (67), or (69) is:

$$P_0 = 1.70\, \theta \qquad (74)$$

Thus we see that if we wish to construct a "perfectly reinforced" gun of *two cylinders* of the same metal, and which will have the most "suitable thickness," so that before firing there shall be developed on the surface of the bore a compressing stress equal to the elastic limit, and so that at the moment of firing the cylinders shall be extended to the elastic limit, then we must fix the full thickness of the wall at 2.28 calibres in which case we will get $1.87\, \theta$ for the elastic strength of the gun.

If we construct a gun under the same conditions, of *three cylinders*, then we must fix the full thickness of the wall at 1.28 calibres, which will give $1.70\, \theta$ for the elastic strength of the gun.

From the preceding, we see that in guns "perfectly reinforced" with reference to stress, as we increase the number of cylinders the thickness of the wall possible for this strengthening diminishes, and the resistance of the gun likewise diminishes. It may therefore be foreseen that with an infinite number of cylinders the thickness of the wall and its elastic strength will be still less if we fulfil the condition of "perfect reinforcement."

It should be recognized that when we speak of "the most advantageous radii," "most suitable thickness of wall," "perfect reinforcement" and "full participation" of all the simple cylinders of a built-up gun, we refer to theoretical conditions which are seldom completely fulfilled in practice; for instance a built-up gun of tube and jacket would never be constructed 2.28 calibres thick, as is indicated in (71), on account of excessive weight of such gun, yet with any other thickness, less or greater, if the elastic limits are equal, the gun cannot be assembled so there will be a "full participation" of both cylinders in the work of resistance.

In this connection the following principle should be understood, that for a given calibre, and fixed number of cylinders composing a gun, that as the *thickness of wall* is diminished the total elastic strength is diminished, but the elastic strength per ton of metal in gun is increased. Also for any given calibre and fixed thickness of wall as we increase the number of cylinders we increase the elastic strength of the gun. These principles find their most complete expression in the wire wound gun, and also explains why a gun composed only of tube and jacket is not made 2.28 calibres thick.

From the preceding discussion it is seen that if we wish to obtain a gun of greater strength than when all cylinders take their full share in the work of resistance, then we must increase the thickness of the wall beyond the deduced thickness, and admit an incomplete participation in the work of resistance, on firing, in some of the exterior cylinders of the gun. At the same time, however, it is impossible to extend the limits of reinforcement indicated in Article 23.

THE LONGITUDINAL RESISTANCE.

28. In a built-up gun the longitudinal stress is borne by the cylinder which carries the breech-block.

In France this is generally the tube; in the United States the cylinder next to the inner tube, that is the jacket.

Let us denote by:

P_0, interior pressure in bore of gun.

R_0, radius of bore (or chamber) of gun.

R_n, R_{n-1}, limiting radii of block carrying cylinder.

The total pressure in the direction of the axis due to the action of the powder gas on the base of the bore will be $\pi P_0 R_0^2$, and the resistance to rupture is equal to the longitudinal tension q

ELEMENTS OF ELASTIC STRENGTH OF GUNS.   49

called into action over the annular cross section $\pi(R_n^2-R_{n-1}^2)$, or:

$$\pi P_0 R_0^2 = q \pi (R_n^2 - R_{n-1}^2)$$

$$\therefore q = \frac{P_0 R_0^2}{R_n^2 - R_{n-1}^2} \qquad (75)$$

In this expression the pressure of the atmosphere has been regarded as zero, and this expression is only true so long as there is no recoil, and then only true for the part of the gun in rear of the trunnions. When the tube carries the breech-block, $n=1$; when the jacket, $n=2$.

FIRING PRESSURES.

29. The built-up gun may be considered in two states; in both of which the pressures act normally.

1°. The state of firing which means the maximum internal pressure acts.

2°. The state of rest which means the internal pressure is 0.

In the state of firing, each cylinder except the outside one, is subjected to the action of the two pressures, one internal and the other external and the outside one is under the action of an internal one only, as the external pressure of the atmosphere is regarded in the state of rest, as 0.

In the state of firing, a longitudinal tension exists which however will be neglected in considering the equilibrium of the forces acting in direction perpendicular to the axis of the cylinder.

Its value is given in (75), and in the U. S. Artillery the cylinder which carries the breech-block has a thicker wall than indicated by the rule for the most advantageous radii; this excess in thickness provides for the longitudinal tension.

In the state of rest the only pressures acting are the pressures due to the assemblage by shrinkage and the inner cylinder (tube) is in the condition of a simple cylinder subjected to external pressure only. The outer cylinder is in the condition of a simple cylinder acted upon by internal pressure only (the pressure of the atmosphere being regarded as 0); each intermediate cylinder however is subjected to the action of two pressures; one internal and the other external.

50   ELEMENTS OF ELASTIC STRENGTH OF GUNS.

The elastic strength of a built-up gun depends upon the radii and the elastic limits of the simple cylinders which compose it.

The condition of any simple cylinder in a built-up gun, either in the state of firing or the state of rest, is determined by the condition of its inner surface; its elastic strength is not exceeded so long as the maximum stress at the inner surface does not exceed the elastic limit in either direction.

In the assembled gun two conditions must be fulfilled: 1°, in the state of firing, the maximum stresses at the interior surfaces of the simple cylinders shall not exceed the elastic limits; 2°, in the state of rest, the compressing stress developed at the surface of the bore due to the pressure of shrinkage, P'_1, shall not exceed p, the elastic limit for compression.

Therefore the *elastic strength* of the built-up gun is deduced from two sets of equations; one of which expresses the condition that in the state of firing the tangential stresses at the interior surfaces of the simple cylinders shall reach their elastic limits; the other expresses the condition that before firing, the tangential stress of compression at the surface of the bore is equal to the elastic limit, and in the state of firing the tangential stress of extension at the surface of the bore is equal to the elastic limit.

Let us denote by:

n, the number of simple cylinders in gun, counting from bore outward.

$R_0, R_1 \ldots R_n$, radii of simple cylinders in gun.

$\theta_0, \theta_1 \ldots \theta_{n-1}$, elastic limits for extension.

$p_0, p_1 \ldots p_{n-1}$, elastic limits for compression.

P_0, pressure in the bore at the instant of firing.

$P_1, P_2, \ldots P_{n-1}$, pressures of firing at shrinkage surfaces.

P''_1, pressure at rest at first shrinkage surface.

Considering the state of firing alone, we will have the tangential elastic strength of the tube from (39); by extending the notation of (39) to the jacket and outer cylinders in succession which compose the built-up gun, we will have formulas giving their elastic strength. Collecting them and beginning with the outside cylinder we will have:

ELEMENTS OF ELASTIC STRENGTH OF GUNS. 51

$$\left.\begin{array}{l} P_{n-1} = \theta_{n-1} \dfrac{R_n^2 - R_{n-1}^2}{R_n^2 + R_{n-1}^2} \qquad\qquad\qquad\qquad\qquad n \\[6pt] P_{n-2} = \theta_{n-2} \dfrac{R_{n-1}^2 - R_{n-2}^2}{R_{n-1}^2 + R_{n-2}^2} + P_{n-1} \dfrac{2 R_{n-1}^2}{R_{n-1}^2 + R_{n-2}^2} \quad n-1 \\[6pt] \cdots\cdots\cdots\cdots\cdots\cdots\cdots\cdots \\[2pt] \cdots\cdots\cdots\cdots\cdots\cdots\cdots\cdots \\[6pt] P_2 = \theta_2 \dfrac{R_3^2 - R_2^2}{R_3^2 + R_2^2} + P_3 \dfrac{2 R_3^2}{R_3^2 + R_2^2} \qquad\qquad\qquad c \\[6pt] P_1 = \theta_1 \dfrac{R_2^2 - R_1^2}{R_2^2 + R_1^2} + P_2 \dfrac{2 R_2^2}{R_2^2 + R_1^2} \qquad\qquad\qquad b \\[6pt] P_0 = \theta_0 \dfrac{R_1^2 - R_0^2}{R_1^2 + R_0^2} + P_1 \dfrac{2 R_1^2}{R_1^2 + R_0^2} \qquad\qquad\qquad a \end{array}\right\} \quad (76)$$

These equations express the condition that in the state of firing, each cylinder is extended to its elastic limit.

The formula which gives the elastic strength of the built-up gun with the condition that in the state of rest the bore is compressed to its elastic limit is (56):

$$P_0 = (\rho_0 + \theta_0) \dfrac{R_n^2 - R_0^2}{R_n^2 + R_0^2} \qquad (77)$$

Hence to determine the elastic strength of a built-up gun we solve (76) and (77); of these two values of P_0, the less will be the maximum permissible firing pressure.

When P_0 from (76) is the greater value, it would be a *safe firing pressure*, but in the state of rest the surface of the bore would be compressed beyond the elastic limit by the pressure of shrinkage at the first shrinkage surface.

When P_0 from (77) is the greater value, so *strong* a shrinkage is required to compress the inner tube in the state of rest to its elastic, limit that the use of P_0 from (77) would extend the interior surface of one or more cylinders, exterior to the tube, beyond the elastic limit.

Whenever two values are found for the elastic strength of a gun, if they are not equal, the less must be the one adopted since the greater is certain to exceed the elastic strength of the gun either in the state of firing or in the state of rest at the inner surface of one cylinder at least.

First, to solve (76) we substitute in (n, 76) the value of θ_{n-1} the resulting value of P_{n-1} will give the elastic strength of the nth cylinder (23).

Substituting this value of P_{n-1} in ($n-1$, 76), P_{n-2} will give the elastic strength of the compound tube formed by the assemblage of the two exterior cylinders.

Continuing this operation and finally substituting the deduced value of P_1 and θ_0 in (a, 76) the resulting value of P_0 will be the maximum value of P_0 if the state of firing be *alone* considered. This value of P_0 will, in connection with the deduced values of $P_1, P_2 \ldots P_{n-1}$, from (76), extend tangentially all cylinders to their elastic limits.

Next solve (77) this value of P_0 gives the firing pressure which will extend the built-up gun regarded as a homogeneous tube, from the limit of compression, ρ_0 at the surface of the bore, to the limit of extension, θ_0.

When P_0 from (76) is the less value we use the firing pressures $P_1, P_2 \ldots P_{n-1}$ deduced from those equations.

When P_0 from (77) is the less value we use it as the elastic strength of the gun (firing pressure.) We have next to find the pressures of firing $P_1, P_2 \ldots P_{n-1}$ corresponding to this value of P_0.

Let us solve (a, 76) with respect to P_1; (b, 76) with respect to P_2 and continue the operation. We will then have:

$$\left. \begin{array}{lr} P_1 = P_0 \dfrac{R_1^2 + R_0^2}{2R_1^2} - \theta_0 \dfrac{R_1^2 - R_0^2}{2R_1^2} & a \\[6pt] P_2 = P_1 \dfrac{R_2^2 + R_1^2}{2R_2^2} - \theta_1 \dfrac{R_2^2 - R_1^2}{2R_2^2} & b \\[4pt] \cdots \cdots \cdots \cdots \cdots \cdots & \\ \cdots \cdots \cdots \cdots \cdots \cdots & \\ P_{n-1} = P_{n-2} \dfrac{R_{n-1}^2 + R_{n-2}^2}{2R_{n-1}^2} - \theta_{n-2} \dfrac{R_{n-1}^2 - R_{n-2}^2}{2R_{n-1}^2} & n-1 \end{array} \right\} \quad (78)$$

In (a, 78) let us substitute P_0 from (77) and deduce P_1; when P_0 and P_1 have these values, the surface of the bore, in the

ELEMENTS OF ELASTIC STRENGTH OF GUNS. 53

state of firing, will be extended tangentially to its elastic limit and when the gun is at rest the pressure at rest, P'_1, due to the assemblage by the computed shrinkage, will compress the surface of the bore tangentially to its elastic limit. In the state of rest there is no danger that any cylinder will be over-compressed except the innermost; we may therefore use the values of P_2, $P_3 \ldots P_{n-1}$ deduced from (76), or we may find a new set of values for P_2, P_3, ... P_{n-1}, from (78). These latter values if used would give the least permissible firing pressures at shrinkage surfaces and also provide for the largest reserve of tangential elastic strength in the outer cylinders. Any convenient value for P_2 may be assigned between the two deduced values from (c, 76) and (b, 78) and the same principle holds for the two deduced values of P_3, P_4 ... P_{n-1}. Means of the two values of P_2, P_3 ... P_{n-1} will probably give the best condition for the stresses in the state of firing, and these mean values would provide for a reserve of tangential strength in the outer cylinders, as will be shown later in "Illustration of Advantages of Shrinkage," Article 31.

SHRINKAGE FORMULAS USING FIRING PRESSURES.

30. The object of assembling a gun by shrinkage is to have all the simple cylinders which compose the built-up gun take their full share in the general resistance of the system, so that in firing, each simple cylinder shall exert its full elastic strength or such less limit of strength as may be intended by the design.

Absolute Shrinkage is the difference between the outer diameter of an inner cylinder and the inner diameter of the cylinder to be assembled upon it.

Relative Shrinkage is the absolute shrinkage divided by the diameter of the shrinkage surface, or is the shrinkage per linear inch, but change per linear inch, by definition of Article 3, is *strain*; hence in order to determine the relative shrinkage at any shrinkage surface we will find the strains in the two cylinders at the surface considered.

At any shrinkage surface let us denote the exterior radius of the inner cylinder by r, and the inner radius of the outer cylinder by R. Representing the difference in diameter by S, we have for the absolute shrinkage:

$$S = 2(r - R)$$

After assemblage, since there is a common surface of contact, the following algebraic relation must obtain whether the gun be in the state of firing or the state of rest:

$$r + \Delta r = R + \Delta R$$
$$\therefore r - R = \Delta R - \Delta r$$

The absolute shrinkage will be:

$$S = 2(\Delta R - \Delta r) \qquad (79)$$

After assemblage the two surfaces whose radii were r and R are in contact and have a common radius, which, compared with the dimensions of the simple cylinders, differs from r and R by so insignificant a quantity that the difference may be neglected and $2R$ taken as the common diameter, then we will have for the relative shrinkage from definition:

$$\varphi_1 = \frac{\Delta R}{R} - \frac{\Delta r}{R} \qquad (80)$$

$\frac{\Delta R}{R}$ represents the change in linear inch of the inner diameter of the outer cylinder; this *strain* is equal to the strain in the circumference; the tangential stress at the inner surface of the outer cylinder is T_1; therefore by applying the third of equation (1), page 6, we have:

$$\frac{\Delta R}{R} = \frac{T_1}{E}$$

In a similar way we deduce

$$\frac{\Delta r}{R} = \frac{T_0'}{E}$$

Hence we see that the tangential strain, $\frac{T_1}{E}$ is equal to the strain in the inner diameter of the outer cylinder, and the tangential strain, $\frac{T_0'}{E}$ is equal to the strain in the outer diameter of the inner cylinder.

Equation (80) shows that the relative shrinkage at any shrinkage surface is the *algebraic* difference between the strains in the two diameters for the surface considered, and moreover that this relation is true whether the gun is in the state of firing or in the state of rest.

We will consider the gun assembled and in the state of firing.

We will denote by:

E, common modulus of elasticity.

$\varphi_1, \varphi_2, \ldots$ relative shrinkages at the surfaces whose radii are respectively $R_1, R_2 \ldots$

$S_1, S_2 \ldots$ absolute shrinkage corresponding to the relative shrinkages $\varphi_1, \varphi_2 \ldots$

Δ_0, absolute contraction of diameter of bore due to the assemblage by shrinkage of all the simple cylinders which make up the gun.

1°. Let us consider a built-up gun of two cylinders (tube and jacket). There will be but one shrinkage surface and we will have for the relative shrinkage after substituting in (80) for the strains, their values as given in third equation of (1), page 6:

$$\varphi_1 = \frac{T_1}{E} - \frac{T'_0}{E} \tag{81}$$

Since in this case $P_2 = 0$, we obtain T_1 by extending the notation of (20) and thus have:

$$T_1 = P_1 \frac{R_2^2 + R_1^2}{R_2^2 - R_1^2}$$

T'_0 produced by the action of P_0 and P_1 is given by (37):

$$T'_0 = P_0 \frac{2R_0^2}{R_1^2 - R_0^2} - P_1 \frac{R_1^2 + R_0^2}{R_1^2 - R_0^2}$$

Making these substitutions in (81) and reducing, we have:

$$\varphi_1 = \frac{P_1}{E} \cdot \frac{2R_1^2 (R_2^2 - R_0^2)}{(R_1^2 - R_0^2)(R_2^2 - R_1^2)} - \frac{P_0}{E} \cdot \frac{2R_0^2}{R_1^2 - R_0^2} \tag{82}$$

The absolute shrinkage will be:

$$S_1 = 2R_1 \varphi_1 \tag{83}$$

The compressing stress at the surface of the bore in the state of rest deduced from (57) is given by:

$$(-T_0) = P_0 \frac{R_2^2 + R_0^2}{R_2^2 - R_0^2} - \theta_0 \tag{84}$$

We deduce for the absolute contraction of bore:

$$\triangle_0 = 2 R_0 \frac{(-T_0)}{E} \tag{85}$$

2°. Let us consider the case of a built-up gun of tube, jacket and hoop, and in the state of firing; there will be two shrinkage surfaces; one between tube and jacket; the other between jacket and hoop.

At the shrinkage surfaces we will have:

$$\varphi_1 = \frac{T_1}{E} - \frac{T'_0}{E} \tag{86}$$

$$\varphi_2 = \frac{T_2}{E} - \frac{T'_1}{E} \tag{87}$$

The value of T_1 depends upon the values of P_1 and P_2 and obtained by extending notation of (36) is given by:

$$T_1 = P_1 \frac{R_2^2 + R_1^2}{R_2^2 - R_1^2} - P_2 \frac{2 R_2^2}{R_2^2 - R_1^2}$$

The value of T'_0 depends on the values of P_0 and P_1 and is given by (37):

$$T'_0 = P_0 \frac{2 R_0^2}{R_1^2 - R_0^2} - P_1 \frac{R_1^2 + R_0^2}{R_1^2 - R_0^2}$$

Since $P_3 = 0$, the value of T_2, obtained by extending notation of (20), is:

$$T_2 = P_2 \frac{R_3^2 + R_2^2}{R_3^2 - R_1^2}$$

The value of T'_1 depends on the values of P_0 and P_2* and obtained by extending notation of (37), is:

$$T'_1 = P_0 \frac{2 R_0^2}{R_2^2 - R_0^2} - P_2 \frac{R_2^2 + R_0^2}{R_2^2 - R_0^2}$$

---

* When two cylinders are to be assembled it is evident, since absolute shrinkage is the difference in diameters *before* assemblage, that the shrinkage will be the same whether the inner cylinder be a simple tube or a compound tube; or the inner cylinder may be regarded as a homogeneous tube without initial tension so far as the determination of the value of shrinkage is concerned.

Substituting for T_1 and T'_0 their values in (86) and for T_2 and T'_1 their values in (87) we obtain after reduction:

$$\varphi_1 = \frac{P_1}{E} \cdot \frac{2R_1^2(R_2^2-R_0^2)}{(R_1^2-R_0^2)(R_2^2-R_1^2)} - \frac{P_0}{E} \cdot \frac{2R_0^2}{R_1^2-R_0^2} - \frac{P_2}{E} \cdot \frac{2R_2^2}{R_2^2-R_1^2} \quad (88)$$

$$\varphi_2 = \frac{P_2}{E} \cdot \frac{2R_2^2(R_3^2-R_0^2)}{(R_2^2-R_0^2)(R_3^2-R_2^2)} - \frac{P_0}{E} \cdot \frac{2R_0^2}{R_2^2-R_0^2} \quad (89)$$

The absolute shrinkages will be

$$S_1 = 2R_1\varphi_1 \quad (90)$$
$$S_2 = 2R_2\varphi_2 \quad (91)$$

The compressing stress at the surface of the bore in the state of rest is deduced from (57):

$$(-T_0) = P_0 \frac{R_3^2+R_0^2}{R_3^2-R_0^2} - \theta_0 \quad (92)$$

The absolute contraction of diameter of bore will be given by:

$$\triangle_0 = 2R_0 \frac{(-T_0)}{E} \quad (93)$$

We will not here deduce shrinkage formulas for a gun containing more than three cylinders. In Part III we will make a more general deduction and the shrinkage formulas in Part II and Part III are identical in form. They only differ in application by the numerical values deduced for the pressures.

It is evident we will have shrinkage formulas, using pressures for the state of rest, if we make the following substitutions in (88) and (89); o for P_0, P'_1 for P_1, and P'_2 for P_2. The pressures at rest are derived from (51) and the method will be fully explained in Part III.

ILLUSTRATION OF THE ADVANTAGES OF SHRINKAGE.

31. The fundamental data which determine the maximum interior pressure a gun can support are its dimensions, its proportions and the elastic strength of the metal used.

In this instance we will assume an 8" gun with thickness of wall equal to one and one-half calibres; also $\theta = \rho = 20$ tons. Any other values would equally well illustrate the theory. We will impose the condition that in the state of firing the inner surfaces of the cylinders composing the gun shall be worked, if practicable, to their elastic limits.

We will take three cases

1° We will find the elastic strength of the homogeneous gun without initial tension (cast gun) with radii, $R_0=4''$ and $R_1=16''$ The maximum interior powder pressure given by (23) is:

$$P_0 = 17.65$$

2°. Next let us consider gun of same dimensions composed of tube and jacket with radii in geometrical progression $R_0=4''$; $R_1=8''$; $R_2=16''$.

P_1 found from (*b*, 76) after substituting $P_2=0$ is:

$$P_1 = 12$$

P_0 obtained from (*a*, 76) is:

$$P_0 = 31.2$$

We do not solve (77) because we know from Article 28 that the surface of the bore in the state of rest will not be compressed to its elastic limit, although the radii are the "most advantageous" yet the thickness is less than the "most suitable".

By dividing 31.2 by 17.65 we see that the gun consisting of two cylinders has 77 per cent. more elastic strength than the simple gun of same dimensions.

3°. We will next consider gun of same dimensions composed of tube, jacket and hoop with radii in geometrical progression $R_0=4''$; $R_1=6''.35$; $R_2=10''.08$; $R_3=16''$.

The radii are the "most advantageous" but the thickness of wall is in excess of 1.23 calibres which was shown in Article 28 to be the "most suitable".

If we fulfil the condition that in the state of firing $T_0=T_1=T_2=20$ tons we know from Article 28 that in the state of rest the surface of the bore will be compressed beyond the elastic limit.

The elastic strength of this gun is therfore given by (77)

$$P_0 = 35.29$$

This is twice the elastic strength of the corresponding homogeneous gun.

In both built-up guns we have assumed the "most advantageous" radii but in neither case the "most suitable thickness"

ELEMENTS OF ELASTIC STRENGTH OF GUNS.   59

to provide for a full participation of all the cylinders in the work of resistance. In the gun of tube and jacket the elastic strength — $P_0 = 31.2$ tons — is less than 1.87θ since the thickness of wall is less than 2.28 calibres.

In the gun of tube, jacket and hoop the elastic strength — $P_0 = 35.29$ tons — is greater than 1.7θ since the thickness of wall is greater than the "most suitable" which was shown to be 1.23 calibres.

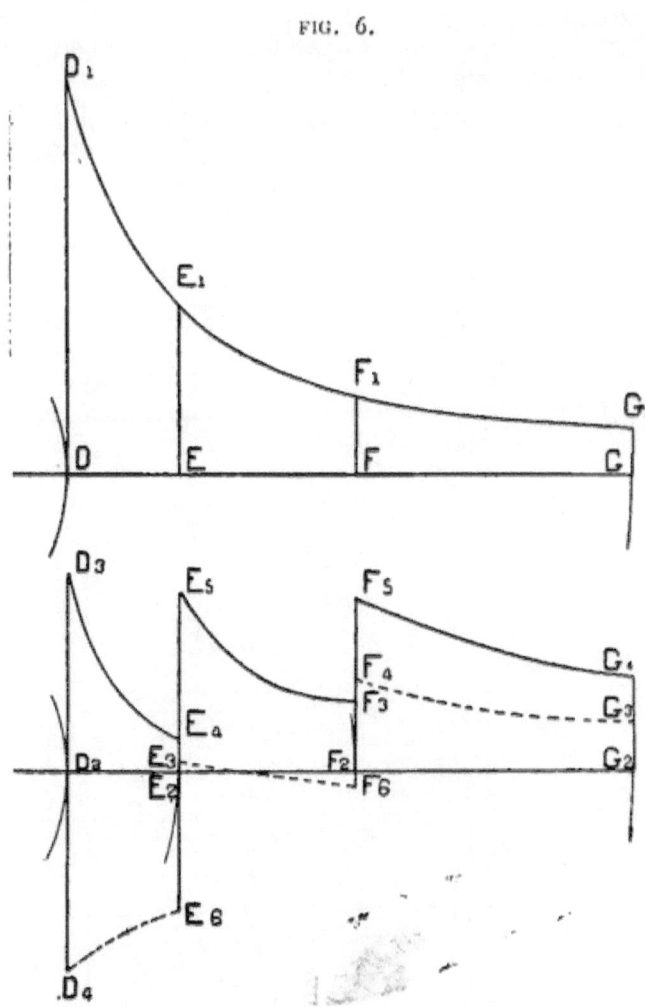

FIG. 6.

In Figure 6 the curve of natural tensions of firing is represented by $D_1 E_1 F_1 G_1$; the curve of tensions of firing in built-up gun by $D_3 E_4$, $E_5 F_8$ and $F_5 G_4$; the curve of tensions at rest by $D_4 E_6$, $E_3 F_6$ and $F_4 G_3$.

The pressures in action, $P_0 = 35.29$, (77); $P_1 = 18.62$, (a, 78); $P_2 = 8.63$, (c, 76); $P_2 = 6.98$ (b, 78); $P_2 = 7.8$ (mean value).

Any value for P_2 greater than 8.63 tons will exceed the elastic strength of the hoop; any value for P_2 less than 6.98 tons will cause the elastic strength of the jacket to be exceeded. As we increase P_2 from 6.98 to 8.63 we diminish the value of T_1 but increase the value of T_2.

Figure 6 has been constructed assuming a mean value of $P_2 = 7.8$. This value of P_2 leaves a reserve of tangential elastic strength in the jacket and hoop, which in each is in excess of 2 tons.

In this illustration the elastic limit for extension in the steel cast gun has been put as high as in the steel forginge of a built-up gun; this is a concession in favor of the cast gun which has never been secured in practice. It is impossible to-day to obtain in an 8″ steel steel cast gun as high a degree of elastic strength as will be found in the separate steel forgings which compose a built-up gun of the same dimensions.

PART III.

PRELIMINARY DISCUSSION.

32. If a rectangular prism be subjected to the simultaneous action of the three stresses, t, p and q in the direction of its edges, the relative strains will be given by (3), but the elastic resistance of the prism will be different depending whether we determine it from the condition that the stresses are equal to the elastic limit or from the condition that the strains or deformations are equal to the relative elongation or contraction at the elastic limit.

Without entering upon a full discussion of the subject we will make the principle clear by a few illustrations.

If the prism be subjected to the tension, $t = \theta$, the strain in the direction of the axis of X will be given by the first of (3), and we get:

$$[t] = \frac{\theta}{E}$$

The stress applied is equal to the elastic limit and the strain corresponds to the elastic limit. We see that any greater stress than θ would permanently deform the prism in the direction of the axis of X. In this case we get the same elastic resistance in the prism whether we consider the stress or the strain as the measure of its strength.

In addition to the tension $t = \theta$, let us apply the tension, $p = \theta$, in the direction of the axis of Y. The strain due to the simultaneous or combined action of t and p will be from the first of (3).

$$[t] = \frac{\theta}{E} - \frac{1}{3} \cdot \frac{\theta}{E} = \frac{2}{3} \cdot \frac{\theta}{E}$$

We see that, although the stresses p and q are each equal to the elastic limit, that the strain in the direction of the axis of X only corresponds to $\frac{2}{3}$ of the elastic limit, or if we let $p = \theta$, we may put $t = \frac{1}{3}\theta$, without $[t]$ being a permanent deformation, or $[t]$ will, under the hypothesis $p = \theta$ and $t = \frac{4}{3}\theta$, be equal to the strain at the elastic limit.

Let us suppose $t = p = q = \theta$, and we will have:

$$[t] = \frac{\theta}{E} - \frac{1}{3} \cdot \frac{\theta}{E} - \frac{1}{3} \cdot \frac{\theta}{E} = \frac{1}{3}\frac{\theta}{E}$$

Or, we see that the prism, subjected to the three stresses t, p and q, each equal to the elastic limit, will only be deformed in the direction of the axis of X by a strain equal to $\frac{1}{3}$ of that produced by the stress $t = \theta$ acting singly, or if we let $p = q = \theta$, we may put $t = \frac{5}{3}\theta$, and $[t]$ will not exceed the elastic limit.

If we suppose $t = p = q = 3\theta$ and acting simultaneously, we will have:

$$[t] = \frac{3\theta}{E} - \frac{\theta}{E} - \frac{\theta}{E} = \frac{\theta}{E}$$

Under the combined effect of all three stresses the prism will not suffer permanent strain, while if either p or q, or both p and q be suppressed, the prism will be permanently deformed.

If, in the first of (3), we put $t = \theta$ and apply p as a pressure, or q as a pressure, or both as pressures, then will $[t]$ be a permanent deformation, however small p or q may be.

Again it is evident that if $t < \theta$ but $> \frac{1}{3}\theta$, and p and q are applied as pressures with values equal to $p = \theta$, then $[t]$ will be a permanent deformation.

Hence it may be said it does not necessarily follow that a prism subjected to stresses will receive a permanent deformation because the applied stress in any direction is equal to, or even in excess of the elastic limit for either extension or compression, but only when the *strain* of *elongation* or *contraction* in any one direction due to the simultaneous effect of all the stresses is greater than that which takes place at the elastic limit under free mechanical tests.

THE COMPOUND ELASTIC STRENGTH OF A TUBE.

33. The elastic strength of a tube as determined by its *resistance to stress*, which may be called its Simple Elastic Strength, (Article 14), we have discussed in Part II of this work, but its

ELEMENTS OF ELASTIC STRENGTH OF GUNS. 63

elastic strength as determined by its *resistance to strain*, which we will call its Compound Elastic Strength, will be discussed in Part III of this text.

The U. S. Army Ordnance Department, in its gun construction, uses "Strain" formulas. A gun (tube) may be said to have reached its limit of elastic strength, as measured by strain, when, from the effect of all the stresses acting upon any portion or element of the gun, both at the moment of firing and before firing, the maximum elastic strains are equal to the limiting strains of elongation or of contraction permissible for the given design and material.

It must not be forgotten that equations (7) and (8) are only true when the tube is a cylinder of revolution of indefinite length, subjected longitudinally to a stress uniformly distributed over the right sections; moreover the strains and stresses must conform to *Hook's* Law; that is, be within the elastic limit.

In the Ordnance Department the formulas of application in artillery engineering, for computing the proper shrinkages for a built-up gun, are deduced under the assumption that the longitudinal stress in a gun, when the system is *in the state of rest*, is 0. The formulas of application are deduced in form to deal with the deformations or strains of elongation or contraction

Substituting in equations (3), $q=0$, we have:

$$\begin{aligned} [t] &= \frac{1}{E}\left(t - \frac{p}{3}\right) \\ [p] &= \frac{1}{E}\left(p - \frac{t}{3}\right) \\ [q] &= \frac{1}{E}\left(-\frac{t}{3} - \frac{p}{3}\right) \end{aligned} \qquad (94)$$

in which t represents the tangential tension due to the resistance of the metal and p the radial pressure; also in which the first members are symbols to represent respectively the strains, or changes in unit of length, of the circumference, of the thickness, and of any element parallel to the axis of the tube.

Eliminating t and p in (94) by substituting their values from (13) and (14) we will have:

$$[t] = \frac{2(P_0 R_0^2 - P_1 R_1^2)}{3E(R_1^2 - R_0^2)} + \frac{4(P_0 - P_1) R_1^2 R_0^2}{3E(R_1^2 - R_0^2)} \cdot \frac{1}{r^2} \quad * \quad (95)$$

$$[p] = \frac{2(P_0 R_0^2 - P_1 R_1^2)}{3E(R_1^2 - R_0^2)} - \frac{4(P_0 - P_1) R_1^2 R_0^2}{3E(R_1^2 - R_0^2)} \cdot \frac{1}{r^2} \quad (96)$$

$$[q] = -\frac{2(P_0 R_0^2 - P_1 R_1^2)}{3E(R_1^2 - R_0^2)} \quad (97)$$

Since t acts tangentially, $[t]$ is the tangential strain in the circumference whose radius is r; it is therefore equal to the change in unit length of radius r, due to the combined effect in a tangential direction, of stress t and radial pressure p.

Since p acts radially, $[p]$ is the radial strain or change in unit length of the thickness (or radial element) of the tube, at the point considered, due to the combined effect, in a radial direction, of the stresses t and p.

Although $q = 0$, and there is no stress in the direction of the axis, yet there is a longitudinal strain of the rectilinear elements due to the action of t and p which is given by $[q]$ in (97).

The three equations, (95), (96) and (97), give respectively the relative changes (*strains*) in the direction of the circumference, of the radial thickness of the walls and of the length of the tube.

Equation (95) gives the tangential strain at any point of the tube corresponding to any assumed value of r, and (96) gives the radial strain (change in thickness of wall) at any point for any assumed value of r. When, in the transformation of these equations, or of equations derived from them, we pass from the *strain* to the *stress* which would, acting singly, produce an equal *strain*, and it may be difficult to determine from the notation whether we are concerned with the tangential or the radial effect of the pressures P_0 and P_1, we will affect P_0 or P_1 with the suffix (θ), when we are considering a tangential effect, and with the suffix (ρ) when we are considering a radial effect. These suffices will be assumed or dropped as convenient.

Equation (95) is the important formula of Part III, since, with a few exceptions, from it or the equations derived from it, is determined the elastic strength of a gun, as measured by strain.

---

* Equations (95), (96) and (97) are Clavarino's formulas as modified by Capt. Rogers Birnie, Ordnance Department, U. S. A.

[t] may be either an elongation or a contraction. In (14) p is always negative, therefore a pressure, but in (96), [p] may be either an elongation or a contraction although p in its radial effect is always compressing.

We will represent the tangential strains of elongation corresponding to the radii R_0 by [T_0], r by [t], R_1 by [T_0'], and the tangential strains of contraction by [$-T_0$], [$-T_0'$].

We will denote the radial strains corresponding to the same radii if elongations by [P_0], [p], [P_1], etc; of contractions by [$-P_0$], [$-P_1$].*

The symbol for a maximum elastic strain of elongation will be [θ], and the symbol for a maximum elastic strain of contraction will be [ρ].

Substituting in succession R_0 and R_1 for r in (95), we have:

$$[T_0] = \frac{P_0(4R_1^2 + 2R_0^2) - P_1 6R_1^2}{3E(R_1^2 - R_0^2)} \quad (98)$$

$$[T_0'] = \frac{P_0 6R_0^2 - P_1(4R_0^2 + 2R_1^2)}{3E(R_1^2 - R_0^2)} \quad (99)$$

Substituting R_0 for r in (96), we obtain:

$$[P_0] = -\frac{P_0(4R_1^2 - 2R_0^2) - 2P_1 R_1^2}{3E(R_1^2 - R_0^2)} \quad (100)$$

Although the notation of equations from (95) to (100) correspond to a separate tube or to the innermost tube of a built-up gun, yet they are deduced from the general equations (3), (13) and (14), under conditions which will apply to any tube; therefore the notation of the preceding formulas or of any formulas derived from them may be extended to any cylinder whatever its location in a built-up gun.

If in the solution or transformation of any of the foregoing equations we find the tangential strain positive, it means an elongation at the cylindrical surface considered; if [t] is negative the tangential strain is a contraction. Also if the radial strain at any cylindrical surface is positive, it means an elongation, and if negative, a contraction.

---

* The general symbols are [t], [T_0], [T_0'], [p], [P_0], [P_1], and it is only for convenience, when, in the solution of an equation, we find a negative strain, that we use with the strain symbol a negative sign to show a contraction.

DISCUSSION OF "STRAIN" EQUATIONS.

34. In this investigation three cases may arise.

First.—When the exterior pressure is zero, or may be so regarded as compared with the interior pressure.

Second.—When the interior pressure is zero.

Third.—When both interior and exterior pressures act.

First Case: $P_1 = o$.

If the exterior surface is free we may consider the pressure on it of the atmosphere as zero, $P_1 = o$; then substituting this value in (95), (96), (97), (98), (99) and (100) we find:

$P_1 = o$

$$[t] = \frac{P_0}{3E} \cdot \frac{2R_0^2}{R_1^2 - R_0^2} \left(\frac{2R_1^2 + r^2}{r^2} \right) \tag{101}$$

$$[p] = -\frac{P_0}{3E} \cdot \frac{2R_0^2}{R_1^2 - R_0^2} \left(\frac{2R_1^2 - r^2}{r^2} \right) \tag{102}$$

$$[q] = -\frac{P_0}{3E} \cdot \frac{2R_0^2}{R_1^2 - R_0^2} \tag{103}$$

$$[T_0] = \frac{P_0}{3E} \cdot \frac{4R_1^2 + 2R_0^2}{R_1^2 - R_0^2} \tag{104}$$

$$[T_0'] = \frac{P_0}{E} \cdot \frac{2R_0^2}{R_1^2 - R_0^2} \tag{105}$$

$$\frac{\theta}{E} = \frac{P_0}{3E} \cdot \frac{4R_1^2 + 2R_0^2}{R_1^2 - R_0^2} \tag{106}$$

$$P_0 = \frac{3\theta(R_1^2 - R_0^2)}{4R_1^2 + 2R_0^2} \tag{107}$$

$$[-P_0] = -\frac{P_0}{3E} \cdot \frac{4R_1^2 - 2R_0^2}{R_1^2 - R_0^2} \tag{108}$$

$$[-P_1] = -\frac{P_0}{3E} \cdot \frac{2R_0^2}{R_1^2 - R_0^2} \tag{109}$$

$$[T_0'] = [T_0] \frac{3R_0^2}{2R_1^2 + R_0^2} \tag{110}$$

$$R_1 - R_0 = R_0 \left(\sqrt{\frac{3\theta + 2P_0}{3\theta - 4P_0}} - 1 \right) \tag{111}$$

* P_1 by hypothesis is equal to zero, but $[-P_1]$ is a symbol to represent the radial strain at the surface whose radius is R_1; this strain is wholly due to $[T_0']$, and is equal to ½ $[T_0']$ and of opposite sign.

By comparing (101) and (102) we see that $[t]$ for the same value of r is greater that $[p]$; or, that at any point of a tube subjected to an interior pressure only, the tangential strain will be greater than the radial stain; hence in this case we need only investigate the elastic strength of the tube with respect to tangential resistance. $[t]$ is positive for all values of r, hence represents an elongation, or the tube is at all points tangentially extended.

$[t]$ increases as r decreases and has its greatest value at the surface of the bore where $r = R_0$; the value of $[t]$ when $r = R_0$ is given by (104), and from it we may find the tangential strain of elongation at the surface of the bore for a given value of P_0; no value of P_0 is admissible which will give a greater value to $[T_0]$ than $\dfrac{\theta}{E}$, the maximum elastic strain of elongation.

$[T_0]$ is produced by the simultaneous action of P_0 and T_0, and the safety of the tube is assured so long as it does not exceed the maximum elastic strain of elongation.

A tangential strain of elongation at the inner surface of the tube is accompanied by an *equal* strain of elongation in the diameter $2R_0$.

When the strain at the surface of the bore is equal to $\dfrac{\theta}{E}$ it does not mean that there is a tangential tension, $T_0 = \theta$, but it means that the tangential *strain* existing at the inner surface is equal to the strain which the metal of the tube would receive, if in a testing machine, a stress equal to θ were applied to it.

The tangential strain of elongation at the outer surface is given by (105), which is the least value of $[t]$.

If we substitute $\dfrac{\theta}{E}$ for $[T_0]$ in (104), we will obtain (106).

If we solve (106) with respect to P_0 we obtain (107), which gives the elastic strength of a cast gun considered as without initial tension. It is apparent from (107)—though not a direct relation—that the elastic strength of a tube depends on its relative dimensions as to calibre; very little additional strength is gained by a considerable increase in thickness of wall. We find from (107) that tubes which are 0.5, 1, and 1.5 calibres thick will support interior pressures equal, respectively, to 0.5θ, 0.63θ and 0.68θ. These tubes being of equal length, their **weights are to each other as 3, 8 and 15.** That is while the greater tube weighs five times as much as the less, yet there is only a gain of 36 per cent. in elastic strength. When the wall is infinitely thick its elastic strength will be:

$$P_0 = \tfrac{3}{4}\theta.$$

Assuming $\theta = \rho$, equation (107) shows since θ, the maximum admissible value of T_0, is always greater than P_0, that cast guns burst, **not directly from the radial pressure P_0 of the powder gases, but from the tangential strains developed in the walls of the gun by the stresses p and t.**

We see from (102) that $[p]$ is always a contraction. It has its greatest numerical value when $r = R_0$, given by (108), and its least numerical value when $r = R_1$ given by (109).

We see by comparing (103) and (109) that the longitudinal strain and radial strain at the **outer surface are equal**; and also since $[q]$ and $[p]$ are negative that the tube is, when compared with original dimensions, shorter and thinner.

If we eliminate P_0 from (104) and (105) and solve for $[T_0']$ we get (110). This gives $[T_0']$ in terms of $[T_0]$ for any interior pressure.

If the gun is 0.5 calibres thick, $[T_0'] = 0.33\,[T_0]$; if 1 calibre thick, $[T_0'] = 0.16\,[T_0]$; if 1.5 calibres thick, $[T_0'] = 0.09\,[T_0]$.

The thickness of a cast gun is seldom in excess of 1.5 calibres, and with this thickness the interior elementary cylinder contributes more than eleven times as much as the exterior one in the general work of resistance.

In order to find in a cast gun the thickness necessary to resist a known interior pressure, P_0, solve (107) with respect to R_1, subtract R_0 from each member and we have (111).

Second Case. $P_0 = 0$.

35. In the case when the tube is subjected to an exterior pressure only, we will have $P_0 = 0$. With this condition in equations (95), (96), (97), (98), (99) and (100) we find:

ELEMENTS OF ELASTIC STRENGTH OF GUNS.

$$P_0 = 0 \begin{cases} [t] = -\dfrac{P_1}{3E} \cdot \dfrac{2R_1^2}{R_1^2 - R_0^2} \left(\dfrac{2R_0^2 + r^2}{r^2} \right) & (112) \\[6pt] [p] = \dfrac{P_1}{3E} \cdot \dfrac{2R_1^2}{R_1^2 - R_0^2} \left(\dfrac{2R_0^2 - r^2}{r^2} \right) & (113) \\[6pt] [q] = \dfrac{P_1}{3E} \cdot \dfrac{2R_1^2}{R_1^2 - R_0^2} & (114) \\[6pt] [-T_0] = -\dfrac{P_1}{E} \cdot \dfrac{2R_1^2}{R_1^2 - R_0^2} & (115) \\[6pt] [-T_0'] = -\dfrac{P_1}{3E} \cdot \dfrac{4R_0^2 + 2R_1^2}{R_1^2 - R_0^2} & (116) \\[6pt] \dfrac{\rho}{E} = \dfrac{P_1}{E} \cdot \dfrac{2R_1^2}{R_1^2 - R_0^2} & (117) \\[6pt] P_1(0) = \rho \dfrac{R_1^2 - R_0^2}{2R_1^2} \quad \{\text{Same as (31.) Stress eq.}\} & (118) \\[6pt] [P_0] = \dfrac{P_1}{3E} \cdot \dfrac{2R_1^2}{R_1^2 - R_0^2} \ast & (119) \\[6pt] [P_1] = \dfrac{P_1}{3E} \cdot \dfrac{4R_0^2 - 2R_1^2}{R_1^2 - R_0^2} & (120) \\[6pt] [-T_0'] = [-T_0]\dfrac{R_1^2 + 2R_0^2}{3R_1^2} & (121) \\[6pt] R_1 - R_0 = R_0 \left(\sqrt{\dfrac{\rho}{\rho - 2P_1}} - 1 \right) & (122) \end{cases}$$

By comparing (112) and (113) we see that $[t]$ for the same value of r will always be numerically greater than $[p]$; hence, in this case we need only investigate the strength of the tube with respect to tangential resistance.

$[t]$ for all values of r is negative, which means the tube is at all points tangentially compressed. $[t]$ increases numerically as r decreases and has its maximum value when $r = R_0$,

---

* P_0 by hypothesis is equal to zero, but the symbol $[P_0]$ represents the radial strain at the surface whose radius is R_0; this strain is wholly due to $[-T_0]$, and is equal to ⅓ $[-T_0]$ but of opposite sign.

given by (115). The safety of the tube is secured so long as the elastic strength at the inner surface is not exceeded.

A tangential strain of contraction at the inner surface of the tube is accompanied by an *equal* strain of contraction in the diameter $2R_0$.

The strain of contraction at the outer surface is given by (116) and is the least value of $[t]$.

From (115) we may find the tangential strain of contraction at the surface of the bore for a given value P_1; no value of P_1 is admissible which will give a greater value to $[-T_0]$ than $\frac{\rho}{E}$ which is the strain corresponding to the elastic limit for compression. Since $P_0 = 0$, $[-T_0]$ is the strain produced by $-T_0$. If we substitute $\frac{\rho}{E}$ for $[-T_0]$ in (115) we obtain (117). If we solve (117) with respect to P_1 we have (118), which gives the tangential elastic strength of a tube subjected to an exterior pressure only. This formula is of especial importance as it limits the exterior pressure on the tube of a built-up gun in the state of rest.

By comparing (108) and (118) we see, if $\theta = \rho$, that a tube can, within the elastic limit, resist a greater interior than exterior pressure.

The radial strain at the outer and inner surfaces of the tube are given respectively by (119) and (120).

At the inner surface, $[P_0]$ is always an elongation.

At the outer surface, $[P_1]$ is an elongation or a contraction depending on the relative values of R_0 and R_1.

Since $[q]$ is always positive, it means that all the tube is lengthened when compared with original dimensions.

By eliminating P_1 in (115) and (116), solving with respect to $[-T_0']$, we obtain (121), which gives the resistance of the outer elementary cylinder in terms of the inner elementary cylinder.

In order to find the thickness of tube necessary to resist a given exterior pressure P_1, solve (118) with respect to R_1, substract R_0 from each member and we will have (122).

ELEMENTS OF ELASTIC STRENGTH OF GUNS.

Third Case: When P_0 and P_1 both act.

36. In this case the tube will be subjected both to an interior and to an exterior pressure.

We will assume equations (95), (96), (97), (98), (99) and (100), and, to preserve the sequence, will renumber them.

$$[t] = \frac{2(P_0 R_0^2 - P_1 R_1^2)}{3E(R_1^2 - R_0^2)} + \frac{4(P_0 - P_1)R_0^2 R_1^2}{3E(R_1^2 - R_0^2)} \cdot \frac{1}{r^2} \quad (95) \quad (123)$$

$$[p] = \frac{2(P_0 R_0^2 - P_1 R_1^2)}{3E(R_1^2 - R_0^2)} - \frac{4(P_0 - P_1)R_0^2 R_1^2}{3E(R_1^2 - R_0^2)} \cdot \frac{1}{r^2} \quad (96) \quad (124)$$

$$[q] = -\frac{2(P_0 R_0^2 - P_1 R_1^2)}{3E(R_1^2 - R_0^2)} \quad (97) \quad (125)$$

$$[t] > [p] \begin{cases} P_0 R_0^2 > P_1 R_1^2 \begin{cases} [T_0] = \frac{P_0(4R_1^2 + 2R_0^2) - 6P_1 R_1^2}{3E(R_1^2 - R_0^2)} & (98)\ (126) \\ [T_0'] = \frac{6P_0 R_0^2 - P_1(4R_0^2 + 2R_1^2)}{3E(R_1^2 - R_0^2)} & (99)\ (127) \\ P_0(\theta) = \frac{3T_0(R_1^2 - R_0^2) + 6P_1 R_1^2}{4R_1^2 + 2R_0^2} & (128) \\ P_0(\theta) = \frac{3\theta(R_1^2 - R_0^2) + 6P_1 R_1^2}{4R_1^2 + 2R_0^2} & (129) \end{cases} \\ P_0 < P_1 \begin{cases} [-T_0] = -\frac{6P_1 R_1^2 - P_0(4R_1^2 + 2R_0^2)}{3E(R_1^2 - R_0^2)} & (130) \\ [-T_0'] = -\frac{P_1(4R_0^2 + 2R_1^2) - 6P_0 R_0^2}{3E(R_1^2 - R_0^2)} & (131) \\ P_1(\theta) = \frac{P_0(4R_1^2 + 2R_0^2) - 3\rho(R_1^2 - R_0^2)}{6R_1^2} & (132) \end{cases} \end{cases}$$

$$[t] < [p] \begin{cases} P_0 > P_1 \begin{cases} [-P_0] = -\frac{P_0(4R_1^2 - 2R_0^2) - 2P_1 R_1^2}{3E(R_1^2 - R_0^2)} & (133) \end{cases} \\ P_0 R_0^2 < P_1 R_1^2 \begin{cases} P_0(\rho) = -\frac{3\rho(R_1^2 - R_0^2) + 2P_1 R_1^2}{4R_1^2 - 2R_0^2} & (134) \end{cases} \end{cases}$$

$$[t] = [p] \begin{cases} P_0 R_0^2 = P_1 R_1^2 \begin{cases} P_0(\theta) = \tfrac{3}{4}\theta & (135) \\ P_0(\rho) = -\tfrac{3}{4}\rho & (136) \end{cases} \\ P_0 = P_1 \{ P_0 = p = -t = \tfrac{3}{2}\rho & (137) \end{cases}$$

In order to determine the location and value of the greatest elastic strains in a tube subjected to the two pressures P_0 and P_1, we can make the three suppositions:

$$[t] > [p]; \qquad [t] < [p]; \qquad [t] = [p].$$

$$[t] > [p].$$

1°. We see by comparing (123) and (124) that the values of $[t]$ and $[p]$ differ only by the sign of the expression

$$\frac{4}{3E} \cdot \frac{(P_0 - P_1) R_0^2 R_1^2}{R_1^2 - R_0^2} \cdot \frac{1}{r^2}$$

hence, if both terms of $[t]$ are positive or both negative, then $[t] > [p]$ numerically.

If $P_0 R_0^2 > P_1 R_1^2$, which involves the condition $P_0 > P_1$ then will both terms of $[t]$ be positive, and all of the tube will be extended tangentially.

If $P_0 < P_1$, which involves the condition $P_0 R_0^2 < P_1 R_1^2$, then will $[t]$ be negative in both terms, and all of the tube will be compressed tangentially.

When $[t] > [p]$ numerically, we are only concerned with the tangential elastic strength of the tube.

An inspection of (123) shows that $[t]$, whether an elongation or contraction, has its greatest numerical value at the surface of the bore, where $r = R_0$ and diminishes as r increases until it has its least value at the outer surface where $r = R_1$. Hence the condition of the surface of the bore is the test of the safety of the tube.

These values of $[T_0]$, $[T_0']$, $[-T_0]$ and $[-T_0']$ are given by (126), (127), (130) and (131).

The value of $[T_0]$ given by the third of (1) is $\frac{T_0}{E}$; making this substitution in (126) and solving with respect to P_0, we get (128). The greatest admissible value of T_0 is θ; making this substitution in (128), we obtain (129), which gives the elastic strength of a tube, subjected both to an interior and exterior pressure, and when $[t] > [p]$, and all of the tube is extended.

If we wish a greater margin of safety than working the tube to the full elastic limit, we may, by assigning reduced values to θ, obtain values for P_0 less than the maximum safe pressure.

The greatest admissible value of $[-T_0]$ is $\dfrac{p}{E}$ making this substitution in (130) and solving for P_1 we obtain (132), which is the elastic strength of the tube when $[t] > [p]$ and all of the tube is compressed. We may find less values for $P_1(\theta)$ than the maximum by assigning less values to p than the elastic limit for compression.

$[T_0]$ represents the tangential strain at the inner surface of the tube, and whether an elongation or a contraction, is accompanied by an *equal* strain of elongation or contraction in the diameter $2R_0$.

In the same way $[T_0']$ represents the tangential strain at the outer surface of the tube, and this strain, whether an elongation or a contraction, is accompanied by an equal strain of elongation or contraction in the diameter $2R_1$.

$$[t] < [p].$$

2°. The only rational hypothesis which will make $[t] < [p]$ is:

and
$$\left.\begin{array}{c} P_0 > P_1 \\ P_0 R_0^2 < P_1 R_1^2 \end{array}\right\}$$

With this condition both terms of $[p]$ in (124) are negative, while the terms of $[t]$ have contrary signs, or the radial strain at all points of the tube will be greater than the tangential strain, and since $[p]$ is negative, all of the tube is radially compressed. When $[t] < [p]$ we are only concerned with the radial elastic strength of the tube to secure its safety.

We see from (124) that $[p]$ increases numerically as r decreases and has its greatest value when $r = R_0$. Hence the safety of the tube is assured so long as the elastic limit of the tube at the surface of the bore is not exceeded in a radial direction. Since all of the tube is compressed with the maximum radial strain at the surface of the bore, we know the limiting value of this contraction is $\frac{\rho}{E}$. Substituting this value for $[-P_0]$ in (133) we obtain (134), which gives the radial elastic strength of a tube subjected both to an interior and to an exterior pressure and when $[t] < [p]$ numerically.

$$[t] = [p].$$

3°. $[t]$ and $[p]$ are only numerically equal under the supposition that:

$$P_0 R_0^2 = P_1 R_1^2$$
or
$$P_0 = P_0$$

With the first condition we see that the greatest values of $[t]$ and $[p]$ are at the surface of the bore. To obtain the elastic strength of the tube, substitute in (123) and (124), R_0 for r and $\frac{P_0 R_0^2}{R_1^2}$ for P_1, $\frac{\theta}{E}$ for $[t]$, and $\frac{\rho}{E}$ for $[p]$. We will then have (135) and (136).

If in (135) and (136) ρ and θ are unequal, we take the less value of P_0 as the elastic strength of the tube.

If $P_0 = P_1$, then at all points $[t]$ and $[p]$ are numerically equal and given by (137).

Equations (129) and (134) are particularly important. The former gives the tangential elastic strength of a tube subjected to the two pressures P_0 and P_1. The latter, the radial elastic strength under the same conditions.

When we wish to find the elastic strength, as determined by strain, of a tube subjected to an interior and to an exterior pressure, we solve both (129) and (134); of the two values deduced for $P_0(\theta)$ and $P_0(\rho)$, we must use the less. If the $P_0(\theta)$ value be the less, the $P_0(\rho)$ value, while a safe pressure in a radial direction, would work the tube beyond the elastic limit in a tangential direction. If the $P_0(\rho)$ value be the less, the $P_0(\theta)$ value cannot be used; although it would give a safe tangential strain, it would over compress the tube in a radial direction.

The principle just demonstrated is very important, and it must be remembered that when two values may be deduced for the elastic strength of a tube—one in a radial and the other in a tangential direction—the less is the one to be used.

A review of this discussion shows that the greatest strain in numerical value is always at the interior surface of the tube, and therefore the condition of the innermost elementary cylinder with respect to maximum strain, either tangential or radial, is really the criterion of the resistance of the tube; if its elastic strength be not exceeded, the integrity of the tube is assured.

STRAINS AT REST, AND OF FIRING. NATURAL STRAINS OF FIRING.

37. The principles enunciated in Article 22 apply also to strains; whence we have:

The strain of firing is the algebraic sum of the natural strain of firing and the strain at rest.

The strain at rest is the algebraic difference of the strain of firing and the natural strain of firing.

The natural strain of firing is the algebraic difference of the strain of firing and the strain at rest.

THE ELASTIC STRENGTH OF A TUBE WITH INITIAL TENSIONS.
THE NATURE OF REINFORCEMENT.

38. In Articles (22) and (23) we explained the increase of resistance in a tube with initial tensions in the case when the elastic strength of the tube is determined by its resistance to *stress*. These initial tensions may be produced by shrinkage, as in a compound tube, (Article 22); by the method of manufacture, as in a gun made by the Rodman process, or by any means which will compress the inner elementary cylinders of the gun.

When *Strain* is taken as the measure of the strength of a tube, simple or compound, similar methods apply.

A built-up gun is one consisting of a series of concentric simple cylinders in which the principle of initial tension is secured by the method of assemblage; by shrinkage as in one composed of cylinders of definite proportions, or by tension of winding as in a wire-wound gun.

The object to be attained is to so dispose the strains upon the cylinders in the state of rest, that the elastic limit for compression of the inner cylinder shall not be exceeded and that each and all of the simple cylinders shall be simultaneously worked to their elastic limit, when the maximum interior pressure, which the system will support, is applied. A perfect realization of this principle is seldom or never reached in practice, since in any particular case for a definite number of cylinders the thickness of the entire wall and the relation of the radii must conform to certain laws to attain the best theoretical results, but in the assemblage of a gun these conditions are modified or excluded by other considerations.

But we can approach more or less nearly this perfect condition of strain, under the maximum pressure, as each particular case may admit. It will be apparent that the more nearly we approach the ideal state of strains in action, the greater will be the tangential strength of the gun, since its aggregate resistance must be a maximum when each simple cylinder is worked to the limit of its elastic strength in resisting the interior pressure.

When the gun is properly assembled, the inner cylinders in the state of rest are compressed, and the outer ones extended by the reaction of the inner cylinders, and the gun will, when the interior pressure reaches the maximum, arrive at the desired state in which each simple cylinder is extended to the limit calculated.

We saw from the discussion of (110) how small a share the exterior parts of a simple tube takes in the general resistance of the tube. We know no strength can be put in a system which does not already exist in it; but it is the merit of *Shrinkage* that it utilizes the elastic strength of all of the simple cylinders composing the built-up gun, and without the device of shrinkage, a compound tube, assembled without initial tension, would have only the elastic strength of a homogeneous tube of the same dimensions.

By putting the outer cylinders under initial extension we utilize so much of their elastic strength as would otherwise be lost and this is the source of gain of strength in a built-up gun over one of simple construction, (homogeneous gun without initial tensions).

In what follows we will consider the built-up gun as composed of cylinders of definite proportions.

ELEMENTS OF ELASTIC STRENGTH OF GUNS. 77

This gun, when made of steel, consists in general of a tube extending the whole length of the bore ; over this, from the breech end is shrunk the jacket, extending about two fifths of the tube, and to these in larger calibres are added one or more rows of hoops also shrunk on, and the entire system is closed at the breech by a movable block.

It is evident that a compressed tube can be extended through a greater range without exceeding the elastic limit for elongation than a neutral tube, that is, one without initial tension.

The nature of reinforcement of a tube depends on the principle that if a tube is *compressed* at its inner surface to its elastic limit, or to a less limit, it can, with safety, be *extended* to its elastic limit, that is, through a range of $[\rho] + [\theta]$ or through the less range of $[-T_0] + [\theta]$.

Equation (129) gives the elastic strength of a compound tube subjected at the instant of firing to an exterior pressure P_1 on the inner cylinder (tube), and in which there is developed at the inner surface of the inner tube by the action of P_0 and T_0, the limiting tangential strain of elongation $[\theta]$. In other words (129) expresses the condition that in the state firing, the innermost cylinder of the built-up gun shall be tangentially extended to its elastic limit.

If we examine the two terms of P_0 in (129) we see the first is a maximum since it gives the elastic strength of a simple tube with radii, R_0 and R_1 (107), when subjected only to an interior pressure.

P_0 will then be a maximum when P_1 is a maximum, but the firing pressure P_1 is limited by the maximum admissible pressure P_1' which may exist before firing at the first shrinkage surface. The maximum value of P_1' is given by (118) since any greater pressure at rest would over compress the surface of the bore. The pressure of firing P_1 is equal to the pressure at rest, plus the natural pressure of firing (50). It is the firing pressure P_1 which in (129) determines the elastic strength of a built-up gun. Let us assume that P_1' at rest compresses the tube to its elastic limit, then when P_0 and P_1 both act, the tube is extended at the surface of the bore from the limiting elastic strain of contraction to the limiting elastic strain of elongation.

The value of P_1' found from (118) is the limiting exterior pressure for the state of rest; if we substitute this value of P_1' for P_1 in (129) we obtain after reduction:

$$P_0 = (\rho+\theta)\frac{3(R_1^2 - R_0^2)}{4R_1^2 + 2R_0^2} \qquad (138)$$

This expression is independent of the exterior pressure. As applied to gun construction, it gives the elastic strength of a simple tube, (107), with thickness of wall equal to $R_1 - R_0$, in which before firing there is, at the surface of the bore, a tangential strain of contraction equal to $[\rho]$ and in which in the state of firing, there is developed the tangential strain of elongation, $[\theta]$; or in other words this formula shows if a simple tube in a state of rest be *compressed* tangentially to its elastic limit at the surface of the bore, that the interior pressure P_0, which develops in this tube at its inner surface a tangential strain equal to the elastic limit, will produce in a simple tube of the same dimensions without initial tension, a tangential strain of elongation equal to $[\rho] + [\theta]$.

If $\rho = \theta$, we see that the elastic strength of the tube given by (138) is twice as great as the elastic strength given by (107).

A gun may be said to be "perfectly reinforced" when, before firing, $[-T_0] = [\rho]$, and at the instant of firing, $[T_0] = [\theta]$.

If $R_1 = \infty$, then will the gun be infinitely thick and:

$$P_0 = \tfrac{3}{4}(\rho+\theta) \qquad (139)$$

which is the limit of the strength of a "perfectly reinforced" gun.

Equation (138) gives the elastic strength of a "naturally hooped" gun, where the initial compression is secured as in the Rodman process by casting hollow and cooling from the interior; or where the initial tension is obtained, as in the Uchatius phosphor-bronze guns, by mandrelling.

It should be stated that the production of initial tension by the methods indicated has not been wholly satisfactory, or in other words it would not be proper to substitute the value of ρ, as determined by physical tests, in (138) in order to find the elastic strength of such a gun, since the result would give too great an elastic strength.

The pressure of firing P_1 in a "naturally hooped" gun is o.

In a built-up gun the tangential strain of contraction at the surface of the bore in the state of rest is caused by the pressure P_1', but P_1' itself results from the contractile force of the cylinders beyond R_1.

Equation (129) gives the elastic resistance of a built-up gun with thickness of wall *greater* than $R_1 - R_0$; the firing pressure P_1 is equal to the sum of the pressure at rest P_1' due to the contractile force of the cylinders assembled beyond R_1, and of the natural pressure of firing p_1 at R_1, (*b*, 50)

If a built-up gun has been properly assembled by shrinkage it may ~~therefore~~ be treated as a simple homogeneous tube; the theory of reinforcement by shrinkage assumes a strain of contraction at the surface of the bore, which may or may not reach the elastic limit; and the theory further gives to the gun an elastic strength which is the powder pressure which will extend the surface of the bore from the limit of its tangential strain of contraction to the limit of the strain for tangential elongation; this principle may be expressed algebraically by extending the notation of (107) to include the limiting radii of the gun and by substituting $(-T_0)^* + \theta_0$ for θ; or if the strain of contraction corresponds to the elastic limit by substituting $\rho_0 + \theta_0$ for θ.

If the exterior radius is R_2 we will have from (107) for the elastic strength of a "perfectly reinforced" gun composed of the tube and jacket:

$$P_0 = (\rho_0 + \theta_0)\frac{3(R_2^2 - R_0^2)}{4R_2^2 + 2R_0^2} \tag{140}$$

If the exterior radius is R_n the formula becomes by extending the notation of (107):

$$P_0 = (\rho_0 + \theta_0)\frac{3(R_n^2 - R_0^2)}{4R_n^2 + 2R_0^2} \tag{141}$$

---

* The positive value of $(-T_0)$ must be substituted in the formula.

Equation (141) gives the elastic strength of a built-up gun with the condition that *before firing* the surface of the bore is compressed to the elastic limit.

The value of P_1 in (129) is dependent on P_1' and if P_1' compresses the surface of the bore exactly to the elastic limit, then P_0, the elastic strength of the gun, will have the same value in (129) and (141).

If the strain of contraction at the surface of the bore, in the state of rest, is less than $[\mu]$, the elastic strength of the gun will be given by:

$$P_0 = \left\{ (-T_0)^* + \theta_0 \right\} \frac{3(R_n^2 - R_0^2)}{4R_n^2 + 2R_0^2} \qquad (142)$$

When P_1 is a maximum the gun may be assembled so that, in the state of firing, each simple cylinder shall be extended to its elastic limit.

The elastic strength of a built-up gun may, then, be given by two equations; one of which, (129), expresses the condition that in the firing state each cylinder shall exert its full elastic strength for extension; the other, (141), expresses the condition that in the firing state the surface of the bore shall be extended to its elastic limit, but contains also the condition that in the state of rest the surface of the bore shall be compressed to the elastic limit. Later, in Article 44, we will explain which of these is the firing pressure to be selected for the deduction of the shrinkages to be applied in the construction of the gun.

This selected value of P_0 will be the maximum service pressure which might be used in the gun.

THE ELASTIC RESISTANCE OF A BUILT-UP GUN IN TERMS OF THE TANGENTIAL STRAINS.

39. Let us suppose the gun assembled by shrinkage, and to contain n simple cylinders.

Let us represent by:

$R_0, R_1, \ldots R_n$, limiting radii of the simple cylinders.

$P_0, P_1, \ldots P_n$, pressures at the instant of firing corresponding to these radii.

---

* See Note, p. 79.

ELEMENTS OF ELASTIC STRENGTH OF GUNS.   81

$T_0, T_1, \ldots T_{n-1}$, tensions of firing corresponding to the tangential strain developed at the inner surfaces of the cylinders.

P_0 is the firing pressure in the bore.

$P_1, P_2 \ldots P_{n-1}$ are the firing pressures at the shrinkage surfaces, due to the combined action of P_0 and the contractile action of the envelope beyond the particular surface considered.

$P_n = 0$, since it is the pressure of the atmosphere.

Equation (128) gives the elastic resistance of the inner tube in action:

$$P_0 = T_0 \frac{3(R_1^2 - R_0^2)}{4R_1^2 + 2R_0^2} + P_1 \frac{3R_1^2}{2R_1^2 + R_0^2}$$

P_1 acts with equal intensity inward and outward, and considered as acting outward, it gives the elastic resistance of the second cylinder when subjected to the two pressures P_1 and P_2. We may find its value in terms of the second cylinder by extending the notation of (128) and obtain:

$$P_1 = T_1 \frac{3(R_2^2 - R_1^2)}{4R_2^2 + 2R_1^2} + P_2 \frac{3R_2^2}{2R_2^2 + R_1^2}$$

(142

Let us assume $n = 3$; whence $P_3 = 0$.

Finding the value of P_2 in terms of the third cylinder we have:

$$P_2 = T_2 \frac{3(R_3^2 - R_2^2)}{4R_3^2 + 2R_2^2}$$

Eliminating P_2 and P_1 from these equations we obtain:

$$\left.\begin{array}{l} P_0 = T_0 \dfrac{3(R_1^2 - R_0^2)}{4R_1^2 + 2R_0^2} + T_1 \dfrac{3R_1^2}{2R_1^2 + R_0^2} \cdot \dfrac{3(R_2^2 - R_1^2)}{4R_2^2 + 2R_1^2} \\[2mm] + T_2 \dfrac{3R_1^2}{2R_1^2 + R_0^2} \cdot \dfrac{3R_2^2}{2R_2^2 + R_1^2} \cdot \dfrac{3(R_3^2 - R_2^2)}{4R_3^2 + 2R_2^2} \end{array}\right\} \quad (143)$$

The law of formation of the separate terms is evident. If the gun consists of n cylinders the formula will be:

82   ELEMENTS OF ELASTIC STRENGTH OF GUNS.

$$\begin{aligned}
P_0 = T_0 \, &\frac{3(R_1^2-R_0^2)}{4\,R_1^2+2R_0^2} + T_1 \, \frac{3R_1^2}{2R_1^2+R_0^2} \cdot \frac{3(R_2^2-R_1^2)}{4R_2^2+2R_1^2} + \\
+\; T_2 \, &\frac{3R_1^2}{2R_1^2+R_0^2} \cdot \frac{3R_2^2}{2R_2^2+R_1^2} \cdot \frac{3(R_3^2-R_2^2)}{4R_3^2+2R_2^2} + \\
+\; T_3 \, &\frac{3R_1^2}{2R_1^2+R_0^2} \cdots \frac{3R_3^2}{2R_3^2+R_2^2} \cdot \frac{3(R_4^2-R_3^2)}{4R_4^2+2R_3^2} + \\
+\; \cdots & \\
\cdots & \cdots \cdots \cdots \cdots \cdots \cdots \cdots + \\
+\; T_{n-1} \, &\frac{3R_1^2}{2R_1^2+R_0^2} \cdots \frac{3R_{n-1}^2}{2R_{n-1}^2+R_{n-2}^2} \cdot \frac{3(R_n^2-R_{n-1}^2)}{4R_n^2+2R_{n-1}^2}
\end{aligned} \quad (144)$$

which gives the resistance of the gun in terms of the positive tensions T_0, T_1, T_2 . . . and T_{n-1} at the inner surfaces of the separate cylinders. When T_0, T_1 . . . and T_{n-1} are equal to the elastic limit, the value of P_0 will be a maximum.

In the expression for P_0 in (144), the first term is the resistance of the tube without initial compression, the second is the share of the second cylinder, and so on, in the general resistance of the wall to the pressure P_0.

It is evident that we may assign various values to T_0, T_1 . . . T_{n-1} not in excess of the elastic limit of the cylinder considered, and thus obtain admissible values of P_0.

Or, as will appear from Article 44, if the gun is in excess of a definite thickness for the number of cylinders employed—we may obtain the same admissible maximum value for P_0 by assigning values to all the tensions except one, and solving, find the value of that special one.

These values of P_0, for the state of firing are, however, limited by the condition that the pressures in the state of rest due to the shrinkage, shall not over compress the inner surface of any cylinder;

THE MOST ADVANTAGEOUS RADIUS FOR SHRINKAGE SURFACE.

40. Having a gun of definite calibre, thickness and number of cylinders, it is now wished to determine the intermediate radii which will give the maximum resistance when the elastic limits are equal.

Let us assume any two adjacent cylinders and denote by
R, interior radius of the inner cylinder.
r, radius of the surface of contact of the two cylinders.
R', exterior radius of the outer cylinder.
P, firing pressure corresponding to radius R.
p, firing pressure corresponding to radius r.
P', firing pressure corresponding to R'.
θ, elastic limit of the cylinders.

We will assume R and R' constant while r varies.

If the resistance of the two cylinders at the instant of firing is a maximum, the maximum tangential stresses developed will be equal to θ.

Applying (129) and changing the notation to correspond to this case, we have:

$$P = \frac{3}{2}\theta\frac{r^2-R^2}{2r^2+R^2} + p\frac{3r^2}{2r^2+R^2}$$

$$p = \frac{3}{2}\theta\frac{R'^2-r^2}{2R'^2+r^2} + P'\frac{3R'^2}{2R'^2+r^2}$$

Eliminating p and solving for P we obtain:

$$P = \frac{\theta}{2}\frac{\left\{-2r^4+r^2(5R'^2-R^2)-2R^2R'^2\right\}+6P'r^2R'^2}{2r^4+r^2(4R'^2+R^2)+2R^2+R'^2}$$

Differentiating and placing the first differential coefficient with respect to r^2 equal to o, we obtain after reduction:

$$(-r^4+R^2R'^2)(3\theta+2P') = 0.$$

Since the second factor is positive we deduce;

$$r^2 = RR' \qquad (145)$$

Hence if two adjacent cylinders in a built-up gun have the same elastic limit, the most advantageous intermediate radius will be a mean proportional between the extreme radii.

In practice other considerations cause a frequent modification of this rule. The cylinder carrying the breech-block and which therefore sustains the longitudinal strain is generally made of a greater thickness than would follow from the law just enunciated. Also if we take the most advantageous radii, the thickness of the simple cylinders might be so small that it would be difficult to prepare them.

TANGENTIAL ELASTIC STRENGTH OF A BUILT-UP GUN WITH MOST ADVANTAGEOUS RADII AND CONSTANT ELASTIC LIMITS.

41. When all of the cylinders of a built-up gun have a common elastic limit and the radii of the cylinders for a given thickness of wall are the "most advantageous" we can reduce (144) to a much simpler form.

Substitute in (144) for $R_1, R_2 \ldots R_{n-1}$ their values taken from (63) and for $T_0, T_1 \ldots T_{n-1}$ their common value θ, and we have:

$$P_0 = \frac{3}{2} \theta \frac{1-a^2}{2+a^2} \left[1 + \left(\frac{3}{2+a^2}\right) + \left(\frac{3}{2+a^2}\right)^2 + \cdots + \left(\frac{3}{2+a^2}\right)^{n-1} \right]$$

Within the brackets we have a geometrical progression in which $\frac{3}{2+a^2}$ is the ratio; summing and reducing we have:

$$P_0 = \frac{3}{2} \theta \left[\left(\frac{3}{2+a^2}\right)^n - 1 \right]$$

or since (64):

$$a = \left(\frac{R_0}{R_n}\right)^{\frac{1}{n}}$$

We obtain after substitution:

$$P_0 = \frac{3}{2} \theta \left\{ \frac{3^n}{\left[2 + \left(\frac{R_0}{R_n}\right)^{\frac{2}{n}}\right]^n} - 1 \right\} \quad (146)$$

THE MOST SUITABLE THICKNESS OF WALL OF A PERFECTLY REINFORCED GUN, WITH CONSTANT ELASTIC LIMITS.

42. If we suppose the gun composed of n cylinders, all having the same elastic limit and conforming to the law of the most advantageous radii, then the elastic strength of the gun is determined from (146) which is deduced under the hypothesis that at the moment of firing, all the cylinders are worked to their elastic limit:

Placing:

$$v = \left(\frac{R_0}{R_n}\right)^{-\frac{2}{n}} = a^2 \quad (147)$$

ELEMENTS OF ELASTIC STRENGTH OF GUNS. 85

we deduce from (146):

$$P_0 = \frac{3}{2}\theta \left\{ \frac{3^n}{(2+v)^n} - 1 \right\} \qquad (148)$$

In this, v fixes the thickness of the wall in calibres.

The elastic strength of the gun determined from the condition that at rest the surface of the bore is compressed to the elastic limit, and in firing the same surface is extended to the elastic limit, is given by (141):

$$P_0 = (p+\theta)\frac{3(R_n^2 - R_0^2)}{4R_n^2 + 2R_0^2}$$

Placing $p = 0$, and observing that

$$R_n^2 = \frac{R_0^2}{v^n} \qquad (149)$$

we get:

$$P_0 = 3\theta\frac{1-v^n}{2+v^n} \qquad (150)$$

Equating (148) and (150), we find:

$$(2+v)^n(4-v^n) = 3^n(2+v^n) \qquad (151)$$

From (151), having assumed the number n of cylinders, we can determine v, which enables us to find the exterior radius R_n from (149). This value of R_n with the known value of R_0 substituted in (148) and (150) will satisfy both. Therefore this value of R_n satisfies, first, the condition that in the state of firing, all the cylinders are extended to their elastic limit and thus perform their full share of the work; second, that in the state of rest the surface of the bore is compressed to the elastic limit.

If R_n is greater than the value deduced from (149) and we compel all the cylinders in firing to exert their full elastic strength, then in a state of rest, the surface of the bore will be compressed beyond the elastic limit.

On the other hand, if at rest the surface of the bore is compressed to the elastic limit then in the state of firing, all the cylinders will not be extended to their elastic limit, or the full elastic strength of all the cylinders will not be exerted.

If R_a is less than the value deduced from (149) and it is wished to have all the cylinders exert their full elastic strength in the state of firing, then before firing, the surface of the bore will not be compressed to its full elastic limit; on the other hand, if the surface of the bore in a state of rest is compressed to its elastic limit, then at the instant of firing, if the inner tube be worked to its elastic limit, the other cylinders will be extended beyond their elastic limit.

R_a deduced from (149) determines the thickness of wall, which is necessary in order that in a state of rest the surface of the bore may be compressed to its elastic limit and in the state of firing all of the cylinders shall be worked to their elastic limit if the gun be properly assembled.

Of the n roots (151) we see by inspection that one is always equal to unity, but it is apparent from (147) that this value cannot be used since R_a is never equal to R_0.

Since R_a is greater than R_0, no value of v can be used except a real positive value less than unity. Assuming $n = 2$, we find that:

$$v = 0.14$$

whence:

$$R_2 = \frac{R_0}{0.14} = 7.1 \, R_0$$

The thickness of wall in calibres will be:

$$\frac{7.1 \, R_0 - R_0}{2 R_0} = 3.05 \qquad (152)$$

The elastic strength of the gun from either (148) or (150) will be:

$$P_0 = 1.45 \, \theta \qquad (153)$$

When $n = 3$ we find:

$$v = 0.42 \therefore v^{\frac{3}{2}} = 0.27$$

and

ELEMENTS OF ELASTIC STRENGTH OF GUNS.   87

The thickness of the wall in calibres will be:

$$\frac{3.7 R_0 - R_0}{2 R_0} = 1.35 \qquad (154)$$

And the elastic strength of the gun by (148) or (150) will be:

$$P_0 = 1.33\,\theta \qquad (155)$$

Thus we see that if we wish to construct a "perfectly reinforced" gun of *two cylinders* of the same metal, and which will have the most "suitable thickness," so that before firing there shall be developed on the surface of the bore a strain of compression equal to the elastic limit, and so that at the moment of firing the cylinders shall be extended to the elastic limit, then we must fix the full thickness of the wall at 3.05 calibres, in which case we will get 1.45 θ for the elastic strength of the gun.

If we construct a gun under the same conditions, of *three cylinders*, then we must fix the full thickness of the wall at 1.35 calibres, which will give 1.33 θ for the elastic strength of the gun.

From the preceding, we see that in guns "perfectly reinforced" with reference to strain, as we increase the number of cylinders the thickness of the wall possible for this strengthening diminishes, and the resistance of the gun likewise diminishes. It may therefore be foreseen that with an infinite number of cylinders the thickness of the wall and its elastic strength will be still less if we fulfil the condition of "perfect reinforcement."

It should be recognized that when we speak of "the most advantageous radii," "most suitable thickness of wall," "perfect reinforcement" and "full participation" of all the simple cylinders of a built-up gun, we refer to the theoretical conditions which are seldom completely fulfilled in practice; for instance a built-up gun of tube and jacket would never be constructed 3.05 calibres thick, as is indicated in (152), on account of the excessive weight of such gun, yet with any other thickness, less or greater, if the elastic limits are equal, the gun cannot be assembled so there will be a "full participation" of both cylinders in the work of resistance.

In this connection the following principle should be understood, that for a given calibre, and fixed number of cylinders composing a gun, that as the *thickness of wall* is diminished the *total* elastic strength is diminished, but the elastic strength per ton of metal in gun is increased. Also for any given calibre and fixed thickness of wall as we increase the number of cylinders we increase the elastic strength of the gun. These principles find their most complete expression in the wire wound gun, and also indicate why a gun composed only of tube and jacket is not made 3.05 calibres thick.

From the preceding discussion it is seen that if we wish to obtain a gun of greater strength than when all cylinders take their full share in the work of resistance, then we must increase the thickness of the wall beyond the deduced thickness, and admit an incomplete participation in the work of resistance, on firing, in some of the exterior cylinders of the gun. At the same time, however, it is impossible to extend the limits of reinforcement indicated in Article 38.

THE LONGITUDINAL RESISTANCE.

43. In the United States Ordnance guns of modern type, the longitudinal stresses, when the gun is fired, are borne by the jacket which carries the breech block. The longitudinal strength will depend upon this cylinder, and the strain due to the pressure of the powder gases must not be in excess of the strain corresponding to the elastic limit of the metal.

To make the demonstration apply to the block carrying cylinder in any gun adopt the following notation:

P_0 interior pressure in bore of gun.

P_n, P_{n-1}, exterior and interior pressures acting directly on block-carrying cylinder.

R_0, radius of bore (or chamber) of gun.

R_n, R_{n-1}, radii respectively of exterior and interior surfaces of block carrying cylinder.

θ, elastic limit for extension of the metal.

E, modulus of elasticity of cylinder.

It is assumed the longitudinal stresses and strains will be uniformly distributed over the cross section of the cylinder.

ELEMENTS OF ELASTIC STRENGTH OF GUNS.

The longitudinal strain at any cross section in rear of the trunnions of the block-carrying cylinder is equal to the algebraic sum of the strains produced in this cylinder, by the strain due to the powder pressure P_0 on the breech block, and the strain produced in same cylinder by the pressures P_{n-1} and P_n.

The first strain—see (75)— is given by the expression :

$$\frac{P_0 R_0^2}{E(R_n^2 - R_{n-1}^2)}$$

and the second strain by exending the notation of (97) is:

$$-\frac{2(P_{n-1} R_{n-1}^2 - P_n R_n^2)}{3 E(R_n^2 - R_{n-1}^2)}$$

Taking the sum of the strains and passing from the total strain to the corresponding stress and representing the resistance per unit of cross section by L we have :

$$L = \frac{3 P_0 R_0^2 - 2(P_{n-1} R_{n-1}^2 - P_n R_n^2)}{3(R_n^2 - R_{n-1}^2)} \qquad (156)$$

which expresses the longitudinal resistance of the cylinder in terms of the pressure in the bore, and the pressures acting normally on the surfaces of the block carrying cylinder.

If the pressures P_n and P_{n-1}, be applied to the jacket cylinder (the second counting from the interior) we will have, since $n = 2$:

$$L = \frac{3 P_0 R_0^2 - 2(P_1 R_1^2 - P_2 R_2^2)}{3(R_2^2 - R_1^2)} \qquad (157)$$

If the tube carries the block, we will have, since $n = 1$:

$$L = \frac{P_0 R_0 + 2 P_1 R_1^2}{3(R_1^2 - R_0^2)} \qquad (158)$$

If the value of L derived from (156) is less than θ, the elastic limit of the metal will not be exceeded. If $L = \theta$, the metal is in danger of being permanently set ; or if L exceeds θ there will be danger of transverse rupture.

FIRING PRESSURES.

44. The built-up gun may be considered in two states; in both of which the pressures act normally.

1°. The *state of firing* which means the maximum internal pressure acts.

2°. The *state of rest* which means the internal pressure is 0.

In the *state of firing*, each cylinder except the outside one, is subjected to the action of two pressures, one internal and the other external, and the outside one is under the action of an internal one only, since the external pressure of the atmosphere is regarded in the state of rest as 0.

In the *state of firing*, a longitudinal strain exists which however will be neglected in considering the equilibrium of the forces acting in direction perpendicular to the axis of the cylinder, its value is given in (158), and in the U. S. Artillery the cylinder which carries the breech-block has a thicker wall than indicated by the rule for the most advantageous radii; this excess in thickness provides for the longitudinal strain.

In the *state of rest* the only pressures acting are the pressures due to the assemblage by shrinkage, and the inner cylinder is in the condition of a simple cylinder subjected to external pressure only; the outer cylinder is in the condition of a simple cylinder acted upon by internal pressure only (the pressure of the atmosphere being regarded as 0); each intermediate cylinder however is subjected to the action of two pressures; one internal and the other external.

The elastic strength of a built-up gun depends upon the radii and the elastic limits of the simple cylinders which compose it.

The condition of any simple cylinder in a built-up gun, either in the state of firing or the state of rest, is determined by the condition of its inner surface; its elastic strength is not exceeded so long as the maximum strain at the inner surface does not exceed the strain at the elastic limit.

In the assembled gun two conditions must then be fulfilled:
1°, in the state of firing, the maximum strains at the interior surfaces of the simple cylinders shall not exceed the elastic limits, 2°, in the state of rest, the strain of contraction $[-T_0]$

developed at the surface of the bore due to the pressure of shrinkage, P_1' shall not exceed $[\rho]$.

Therefore the *elastic strength* of the built-up gun is deduced from two sets of equations; one of which expresses the condition that in the state of firing the strains at the interior surfaces of the simple cylinders shall reach their elastic limits; the other expresses the condition that before firing, the tangential strain of contraction at the surface of the bore is equal to the elastic limit, and in the state of firing the tangential strain of elongation at the surface of the bore is equal to the elastic limit.

Let us denote by:

n, the number of simple cylinders in gun, counting from bore outward

$R_0, R_1 \ldots R_n$, radii of simple cylinders in gun.

$\theta_0, \theta_1 \ldots \theta_{n-1}$, elastic limits for extension.

$\rho_0, \rho_1 \ldots \rho_{n-1}$, elastic limits for compression.

P_0, pressure in the bore at the instant of firing.

$P_1, P_2 \ldots P_{n-1}$, pressures of firing at shrinkage surfaces.

P_1', pressure at rest at first shrinkage surface.

In deducing Firing Pressures in Part III we use the formulas which give the resistance of the gun to *strain*. The maximum strains are always at the inner surfaces of the simple cylinders composing the built-up gun but they may be either in a tangential or in a radial direction. We have therefore to use for each simple cylinder in the gun, two formulas, one which gives its tangential elastic strength, (129), the other its radial elastic strength, (134).

In the outer cylinder, since for gun steel ρ is never less than θ, the tangential elastic strength is always the less than the radial (Article 34) and the one to be used. For the tube the Ordnance Department, U. S. A., uses the $P_0(\theta)$ value in computing shrinkages whether it be less or greater than the $P_0(\rho)$ value, but the less value of P_0 limits the firing pressure.

For all other cylinders we use the less value for the elastic strength. If the (θ) value is the greater its use will over compress the cylinder in a radial direction; if the (ρ) value is the greater its use will extend tangentially the inner surface of the cylinder beyond the elastic limit.

Therefore the equations for **the elastic strength of each cylinder** will be in pairs and after solving **both, the less value of** P_0 will be used, with the exception before noted that we adopt the $P_0(\theta)$ value for the interior pressure in computation for shrinkage.

In the state of firing, the tangential elastic strength of the tube is given by (129) and the radial elastic strength by (134); extending the notation of these equations to make them apply in sequence to the simple cylinders, we have, beginning with the outer cylinder:

$$\left.\begin{array}{l} P_{n-1}(\theta) = \dfrac{3\theta_{n-1}(R_n^2 - R_{n-1}^2)}{4R_n^2 + 2R_{n-1}^2} \\[6pt] P_{n-2}(\theta) = \dfrac{3\theta_{n-2}(R_{n-1}^2 - R_{n-2}^2) + 6P_{n-1}R_{n-1}^2}{4R_{n-1}^2 + 2R_{n-2}^2} \\[6pt] P_{n-2}(p) = -\dfrac{3\theta_{n-2}(R_{n-1}^2 - R_{n-2}^2) + 2P_{n-1}R_{n-1}^2}{4R_{n-1}^2 - 2R_{n-2}^2} \\[6pt] \quad \cdot\ \cdot\ \cdot\ \cdot\ \cdot\ \cdot\ \cdot\ \cdot\ \cdot \\[2pt] \quad \cdot\ \cdot\ \cdot\ \cdot\ \cdot\ \cdot\ \cdot\ \cdot\ \cdot \\[6pt] P_1(\theta) = \dfrac{3\theta_1(R_2^2 - R_1^2) + 6P_2R_2^2}{4R_2^2 + 2R_1^2} \\[6pt] P_1(p) = -\dfrac{3\theta_1(R_2^2 - R_1^2) + 2P_2R_2^2}{4R_2^2 - 2R_1^2} \\[6pt] P_0(\theta) = \dfrac{3\theta_0(R_1^2 - R_0^2) + 6P_1R_1^2}{4R_1^2 + 2R_0^2} \\[6pt] P_0(p) = -\dfrac{3\theta_0(R_1^2 - R_0^2) + 2P_1R_1^2}{4R_1^2 - 2R_0^2} \end{array}\right\} \quad (159)$$

These equations express the condition that in the state of firing each cylinder is worked to its elastic limit.

The (θ) values give the firing pressures which will extend each separate cylinder to its tangential limit, and the (p) values give the firing pressures which will compress each separate cylinder to its radial limit. For each pair of equations in (159) we must adopt for the firing pressure the less numerical value. We have now a set of values for $P_0, P_1 \cdot \cdot \cdot P_{n-1}$ which will, in the state of firing, work each cylinder of the built-up gun to its elastic limit.

ELEMENTS OF ELASTIC STRENGTH OF GUNS.

The formula which gives the tangential elastic strength of a built-up gun with the condition that in the state of rest the tangential strain of contraction at the surface of the bore corresponds to the strain at the elastic limit is (141).

$$P_0 = (p_0 + \theta_0) \frac{3(R_u^2 - R_0^2)}{4R_u^2 + 2R_0^2} \qquad (160)$$

Hence, to determine the elastic strength of a built-up gun as measured by strain, solve (159) and (160); of these two values of P_0 the less will be the maximum permissible firing pressure.

First, to solve equations (159) if we substitute in (n, 159) the value of θ_{n-1} as determined by physical tests, the resulting value of P_{n-1} will be the elastic strength of the nth cylinder.

Substituting this deduced value of P_{n-1} in the two equations ($n-1$, 159) the less value of P_{n-2} will give the elastic strength of the compound tube formed by the assemblage of the two outer cylinders.

Continuing this operation and finally substituting in (a, 159) θ_0 and the less value of P_1, the resulting value of $P_0(\theta)$ will give the tangential strength of the built-up gun in the state of firing. The values of $P_0(p)$ will also be determined, if it is greater than $P_0(\theta)$, then the latter will maximum safe powder pressure; if $P_0(p)$ is the less value, then after the gun has been assembled using the $P_0(\theta)$ value in our computations for shrinkage no firing pressure will be allowed greater than $P_0(p)$.

These selected values of P_0, P_1 . . . P_{n-1} are the maximum permissible firing pressures if the state of firing be alone considered.

Next solve (160) this value of P_0 gives the firing pressure which will extend the surface of the bore of the built-up gun regarded as a homogeneous structure from the limiting tangential strain of contraction [p_0] to the limiting strain of elongation [θ_0].

When the selected value of P_0 from (159) is greater than P_0 from (160) it is a *safe firing pressure* but in the state of rest the surface of the bore will be compressed tangentially beyond the limiting strain by the pressure of shrinkage at the first shrinkage surface.

94   ELEMENTS OF ELASTIC STRENGTH OF GUNS.

When P_0 from (160) is the greater value, the pressures at rest at the shrinkage surfaces required to compress the surface of the bore to its elastic limit will be so great that this value of P_0 will work all the cylinders, except the innermost, beyond the elastic limit.

When two values are found for the elastic strength of a gun, if they are not equal, the less must be the one adopted, since the greater will certainly exceed the elastic strength of the gun in either the state of firing or in the state of rest and in a tangential or radial direction at the inner surface of one or more cylinders.

When P_0 from (159) is less than the P_0 from (160) we use the selected values of the firing pressures $P_1, P_2 \ldots P_{n-1}$ deduced from (159).

When P_0 from (160) is the less value we use it as the maximum permissible powder pressure. We have next to find firing pressures $P_1, P_2 \ldots P_{n-1}$ corresponding to this value of P_0.

Let us solve (a, 159) with respect to $P_1(\theta)$; (b, 159) with respect to $P_2(\theta)$ and continuing the operation we will have:

$$
\begin{aligned}
P_1(\theta) &= \frac{P_0(4R_1^2 + 2R_0^2) - 3\theta_0(R_1^2 - R_0^2)}{6R_1^2} & a \\
P_2(\theta) &= \frac{P_1(4R_2^2 + 2R_1^2) - 3\theta_1(R_2^2 - R_1^2)}{6R_2^2} & b \\
&\ldots\ldots\ldots\ldots\ldots \\
&\ldots\ldots\ldots\ldots\ldots \\
P_{n-1}(\theta) &= \frac{P_{n-2}(4R_{n-1}^2 + 2R_{n-2}^2) - 3\theta_{n-2}(R_{n-1}^2 - R_{n-2}^2)}{6R_{n-1}^2} & n-1
\end{aligned}
\quad (161)
$$

In (a, 161) let us substitute P_0 from (160) the resulting value of P_1 is the firing pressure which in connection with P_0 will work the tube to its elastic limit for tangential elongation and when the gun is in the state of rest, the pressure at rest P'_1, due to the assemblage of the gun by the computed shrinkage, will compress the surface of the bore to its elastic limit for tangential contraction.

In the state of rest there is no probability any cylinder will be over compressed except the innermost; we may therefore use, if we wish, the values of $P_2, P_3 \ldots P_{n-1}$ deduced from (159) or we may find a new set of values for $P_2, P_3 \ldots P_{n-1}$ from (161). These values might be used and would give the least permissible firing pressures at the **shrinkage** surfaces and also provide for the maximum **reserve of** tangential elastic strength in the outer cylinders. Any convenient **value** for P_2 may be assigned between the two deduced values from (159) and (161) and the same rule holds for the two deduced values of $P_3 \ldots$ and P_{n-1}. Means of the two values of $P_2 \ldots$ and P_{n-1} will probably give the best condition in the state of firing and these mean values will provide for a substantial reserve of elastic strength in the outer cylinders.

It may be of interest to remark that if P_0 from (160) is greater than P_0 from (159) and we deduce from (161) the values of P_1, $P_2 \ldots P_{n-1}$ corresponding to the P_0 from (160) and assemble the gun with shrinkages dependent upon these firing pressures, then in the state of firing the surface of the bore will be worked from the limit of contraction to the limit for elongation but the interior surfaces of all the other cylinders will be extended beyond their elastic limit of elongation.

SHRINKAGE FORMULAS USING FIRING PRESSURES.

45. The object of assembling a gun by shrinkage is to have all the simple cylinders which compose the built-up gun take their full share in the general resistance of the system, so that in firing, each simple cylinder shall exert its full elastic strength or such less limit of strength as may be intended for the particular design.

Absolute shrinkage is the difference between the outer diameter of an inner cylinder and the inner diameter of the cylinder to be assembled upon it.

Relative shrinkage is the absolute shrinkage divided by the diameter of the shrinkage surface, or is the shrinkage per linear inch, but change per linear inch, by definition of Article 3, is strain.

At any shrinkage surface let us denote the outer radius of the inner cylinder by r, and the inner radius of the outer cylinder by R. Representing the difference in diameter by S, we have for the absolute shrinkage:

$$S = 2(r - R)$$

After assemblage, since there is a common surface of contact, the following algebraic relation must obtain whether the gun be in the state of firing or the state of rest, or in the state of partial assemblage:

$$r + \Delta r = R + \Delta R$$
$$\therefore \quad r - R = \Delta R - \Delta r$$

The absolute shrinkage will be:

$$S = 2(\Delta R - \Delta r) \tag{162}$$

After assemblage the two surfaces whose radii were r and R are in contact and have a common radius, which, compared with the dimensions of the simple cylinders, differs from r and R by so insignificant a quantity that the difference may be neglected and $2R$ taken as the common diameter, then denoting the relative shrinkage by φ we will have from definition:

$$\varphi = \frac{\Delta R}{R} - \frac{\Delta r}{R} \tag{163}$$

$\frac{\Delta R}{R}$ represents the change in linear inch of the inner diameter of the outer cylinder; this *strain* is equal to the tangential strain in the inner surface of the same cylinder.

$\frac{\Delta r}{R}$ represents the change in linear inch of the outer diameter of the inner cylinder; this *strain* is equal to the tangential strain in the outer surface of the same cylinder.

Let us consider the mth shrinkage surface and denote the relative shrinkage by φ_m; the strain in the inner diameter of the outer cylinder is equal to the tangential strain $[T_m]$; the strain in the outer diameter of the inner cylinder is equal to $[T'_{m-1}]$.

Making these substitutions in (163) we will have for the mth shrinkage surface:

$$\varphi_m = [T_m] - [T'_{m-1}] \tag{164}$$

Equation (164) shows that the relative shrinkage at any shrinkage surface is the *algebraic* difference between the tangential strains in the two surfaces which after assemblage have a common diameter, and moreover that this relation is true whether the gun is in the state of firing or in the state of rest, or in the state of partial assemblage.

We will consider the gun assembled containing n simple cylinders and in the state of firing.

We will denote by:

E, common modulus of elasticity.

$\varphi_1, \varphi_2, \ldots \varphi_m \ldots \varphi_{n-1}$ relative shrinkages at the surfaces whose radii are respectfully $R_1, R_2, \ldots R_m \ldots R_{n-1}$.

$S_1, S_2, \ldots S_m \ldots S_{n-1}$ absolute shrinkage corresponding to the relative shrinkages $\varphi_1, \varphi_2, \ldots \varphi_m \ldots \varphi_{n-1}$.

$\delta_1, \delta_2 \ldots \delta_m \ldots \delta_{n-1}$ partial contractions of diameter of bore due to successive shrinkages.

Δ_0, absolute contraction of diameter of bore due to the assemblage by shrinkage of all the simple cylinders which make up the gun.

Applying (164) to the first shrinkage surface between tube and jacket we will have:

$$\varphi_1 = [T_1] - [T'_0] \qquad (165)$$

Applied to the second shrinkage surface between jacket and hoop we will have;

$$\varphi_2 = [T_2] - [T'_1] \qquad (166)$$

The general equation will be (164) as above:

$$\varphi_m = [T_m] - [T'_{m-1}] \qquad (167)$$

These values of $\varphi_1, \varphi_2, \ldots \varphi_m$ are deduced for the state of firing and in making the deduction we will consider the radii of the bore, of the shrinkage surface and of the next exterior surface.

For the first shrinkage surface between tube and jacket we will have by applying (126) and (127)

$$[T_1] = \frac{P_1(4R_2^2 + 2R_1^2) - 6P_2 R_2^2}{3E(R_2^2 - R_1^2)} \qquad (168)$$

$$[T'_0] = \frac{6P_0 R_0^2 - P_1(4R_0^2 + 2R_1^2)}{3E(R_1^2 - R_0^2)} \qquad (169)$$

We deduce for the second shrinkage surface applying (126) and (127):

$$[T_2] = \frac{P_2(4R_3^2+2R_2^2)-6P_3R_3^2}{3E(R_3^2-R_2^2)}$$

$$[T'_1] = \frac{6P_0R_0^2-P_2(4R_0^2+2R_2^2)}{3E(R_2^2-R_0^2)}$$

We will have for the general terms:

$$[T_m] = \frac{P_m(4R_{m+1}^2+2R_m^2)-6P_{m+1}R_{m+1}^2}{3E(R_{m+1}^2-R_m^2)}$$

$$[T'_{m-1}] = \frac{6P_0R_0^2-P_m(4R_0^2+2R_m^2)}{3E(R_m^2-R_0^2)}$$

Collecting and reducing we have:

$$\varphi_1 = \frac{2P_1R_1^2(R_2^2-R_0^2)}{E(R_1^2-R_0^2)(R_2^2-R_1^2)} - \frac{2P_0R_0^2}{E(R_1^2-R_0^2)} - \frac{2P_2R_2^2}{E(R_2^2-R_1^2)} \quad (170)$$

$$\varphi_2 = \frac{2P_2R_2^2(R_3^2-R_0^2)}{E(R_2^2-R_0^2)(R_3^2-R_2^2)} - \frac{2P_0R_0^2}{E(R_2^2-R_0^2)} - \frac{2P_3R_3^2}{E(R_3^2-R_2^2)} \quad (171)$$

$$\cdots \cdots \cdots \cdots \cdots$$

$$\varphi_m = \frac{2P_mR_m^2(R_{m+1}^2-R_0^2)}{E(R_m^2-R_0^2)(R_{m+1}^2-R_m^2)} - \frac{2P_0R_0^2}{E(R_m^2-R_0^2)} - \frac{2P_{m+1}R_{m+1}^2}{E(R_{m+1}^2-R_m^2)} \quad (172)$$

Representing the absolute shrinkages by $S_1, S_2 \ldots S_m$, we will have:

$$S_1 = 2R_1\varphi_1 \quad (173)$$

$$S_2 = 2R_2\varphi_2 \quad (174)$$

$$\cdots \cdots$$

$$S_m = 2R_m\varphi_m \quad (175)$$

The compressing stress $(-T_0)$ at the surface of the bore in the state of rest is deduced from (142) and is given by

$$(-T_0) = P_0\frac{4R_n^2+2R_0^2}{3(R_n^2-R_0^2)} - \theta_0 \quad (176)$$

Denoting the absolute contraction of diameter of bore by Δ_0, its value is deduced from

$$\Delta_0 = \frac{2R_0(-T_0)}{E} \quad (177)$$

ELEMENTS OF ELASTIC STRENGTH OF GUNS.

The total contraction of diameter of bore is equal to the sum of the partial contractions of the bore due to the successive shrinkages, or:

$$\Delta_0 = \delta_1 + \delta_2 + \ldots \delta_m + \ldots \delta_{n-1} \quad (178)$$

The relative shrinkage at the mth shrinkage surface, φ_m, has always the same value, it is not affected by the state of firing, by the state of rest or by the shrinking on of subsequent cylinders.

Let us consider the gun after the assemblage of the $(m+1)$th cylinder; in this case we have $P_0 = 0$, and $P_{m+1} = 0$.

Making these substitutions in (172) and multiplying both members by:

$$\frac{R_{m+1}^2 - R_m^2}{R_{m+1}^2 - R_0^2}$$

$$* \left\{ (-T_0) = -\frac{2\,\sigma_{\!i}\,l}{E(R_i^2\ldots)} \text{ when } P_0 = 0 \right.$$

we will have:

$$\varphi_m \frac{R_{m+1}^2 - R_m^2}{R_{m+1}^2 - R_0^2} = \frac{2 P_m R_m^2}{E(R_m^2 - R_0^2)} \quad (179)$$

Comparing the second member of (179) with (115)* we see it gives the value of the tangential strain of contraction at the surface of the bore of a tube whose radii are R_0 and R_m; therefore the strain of contraction of the diameter of bore of this tube in terms of φ_m and *independent of the pressure* is given by the first member of (179).

The absolute partial contraction of diameter of bore due the mth shrinkage will then be:

$$\delta_m = 2 R_0 \varphi_m \frac{R_{m+1}^2 - R_m^2}{R_{m+1}^2 - R_0^2} \quad (180)$$

Hence we will have for verifying the absolute contraction of diameter of bore in terms of the partial contractions (178)

$$\Delta_0 = 2 R_0 \left\{ \varphi_1 \frac{R_2^2 - R_1^2}{R_2^2 - R_0^2} + \cdot \varphi_m \frac{R_{m+1}^2 - R_m^2}{R_{m+1}^2 - R_0^2} + \cdot \varphi_{n-1} \frac{R_n^2 - R_{n-1}^2}{R_n^2 - R_0^2} \right\} \quad (181)$$

1° These formulas for a built-up gun of tube and jacket become:

$$\varphi_1 = \frac{2 P_1 R_1^2 (R_2^2 - R_0^2)}{E(R_1^2 - R_0^2)(R_2^2 - R_1^2)} - \frac{2 P_0 R_0^2}{E(R_1^2 - R_0^2)} \quad (182)$$

$$S_1 = 2 R_1 \varphi_1 \quad (183)$$

$$(-T_0) = P_0 \frac{4R_2^2 + 2R_0^2}{3(R_2^2 - R_0^2)} - \theta_0 \qquad (184)$$

$$\Delta_0 = \frac{2R_0(-T_0)}{E} \qquad (185)$$

2° These formulas become for a built-up gun of tube, jacket and hoop:

$$\varphi_1 = \frac{2P_1 R_1^2 (R_2^2 - R_0^2)}{E(R_1^2 - R_0^2)(R_2^2 - R_1^2)} - \frac{2P_0 R_0^2}{E(R_1^2 - R_0^2)} - \frac{2P_2 R_2^2}{E(R_2^2 - R_1^2)} \qquad (186)$$

$$\varphi_2 = \frac{2P_2 R_2^2 (R_3^2 - R_0^2)}{E(R_2^2 - R_0^2)(R_3^2 - R_2^2)} - \frac{2P_0 R_0^2}{E(R_2^2 - R_0^2)} \qquad (187)$$

$$S_1 = 2R_1 \varphi_1 \qquad (188)$$

$$S_2 = 2R_2 \varphi_2 \qquad (189)$$

$$(-T_0) = P_0 \frac{4R_3^2 + 2R_0^2}{3(R_3^2 - R_0^2)} - \theta_0 \qquad (190)$$

$$\Delta_0 = \frac{2R_0(-T_0)}{E} \qquad (191)$$

The formula for verifying the total contraction of diameter of bore by the partial contractions will be:

$$\Delta_0 = 2R_0 \left\{ \varphi_1 \frac{R_2^2 - R_1^2}{R_2^2 - R_0^2} + \varphi_2 \frac{R_3^2 - R_2^2}{R_3^2 - R_0^2} \right\} \qquad (192)$$

3° These formulas become for a built-up gun of four cylinders:

$$\varphi_1 = \frac{2P_1 R_1^2 (R_2^2 - R_0^2)}{E(R_1^2 - R_0^2)(R_2^2 - R_1^2)} - \frac{2P_0 R_0^2}{E(R_1^2 - R_0^2)} - \frac{2P_2 R_2^2}{E(R_2^2 - R_1^2)} \qquad (193)$$

$$\varphi_2 = \frac{2P_2 R_2^2 (R_3^2 - R_0^2)}{E(R_2^2 - R_0^2)(R_3^2 - R_2^2)} - \frac{2P_0 R_0^2}{E(R_2^2 - R_0^2)} - \frac{2P_3 R_3^2}{E(R_3^2 - R_2^2)} \qquad (194)$$

$$\varphi_3 = \frac{2P_3 R_3^2 (R_4^2 - R_0^2)}{E(R_3^2 - R_0^2)(R_4^2 - R_3^2)} - \frac{2P_0 R_0^2}{E(R_3^2 - R_0^2)} \qquad (195)$$

$$S_1 = 2R_1 \varphi_1 \qquad (196)$$

$$S_2 = 2R_2 \varphi_2 \qquad (197)$$

$$S_3 = 2R_3 \varphi_3 \qquad (198)$$

$$(-T_0) = P_0 \frac{4R_4^2 + 2R_0^2}{3(R_4^2 - R_0^2)} - \theta_0 \qquad (199)$$

$$\Delta_0 = \frac{2R_0(-T_0)}{E} \qquad (200)$$

The formula for verifying the total contraction of diameter of bore by the partial contractions will be:

$$\Delta_0 = 2R_0 \left\{ \varphi_1 \frac{R_2^2 - R_1^2}{R_2^2 - R_0^2} + \varphi_2 \frac{R_3^2 - R_2^2}{R_3^2 - R_0^2} + \varphi_3 \frac{R_4^2 - R_3^2}{R_4^2 - R_0^2} \right\} \qquad (201)$$

PRESSURES AT REST.*

46. To obtain the Pressures at Rest at the several shrinkage surfaces we find; 1° the Firing Pressures $P_0, P_1, P_2 \ldots P_m \ldots P_{n-1}$ as explained in Article 44. 2° The natural Pressures of Firing $p_1, p_2 \ldots p_m \ldots P_{n-1}$ from (19) by substituting R_n for R_1, and then in succession $R_1, R_2, \ldots R_m \ldots R_{n-1}$ for r, thus:

$$\left. \begin{array}{ll} r = R_1; & p_1 = \dfrac{P_0 R_0^2 (R_n^2 - R_1^2)}{R_1^2 (R_n^2 - R_0^2)} \qquad a \\[6pt] r = R_2; & p_2 = \dfrac{P_0 R_0^2 (R_n^2 - R_2^2)}{R_2^2 (R_n^2 - R_0^2)} \qquad b \\[6pt] \cdots \cdots \cdots \cdots \\[6pt] r = R_m; & p_m = \dfrac{P_0 R_0^2 (R_n^2 - R_m^2)}{R_m^2 (R_n^2 - R_0^2)} \qquad m \end{array} \right\} \qquad (202)$$

3° Having obtained from (202) the natural pressures we deduce the pressures at rest from (51).

$$\left. \begin{array}{ll} P'_0 = 0 & a \\ P'_1 = P_1 - p_1 & b \\ P'_2 = P_2 - p_2 & c \\ \cdots \cdots \cdots \\ P'_m = P_m - p_m & m \\ \cdots \cdots \cdots \\ P'_{n-1} = P_{n-1} - p_{n-1} & n-1 \end{array} \right\} \qquad (203)$$

---

* It is interesting to compare the shrinkage formulas in Articles 30 and 46 and to note that they are identical in form; in application they differ by the values deduced for the Firing Pressures, $P_0, P_1 \ldots P_m \ldots P_{n-1}$.

If the value of P_0 in (202) is derived from (159), then the surface of the bore of the assembled gun in the state of rest will not be tangentially compressed to its elastic limit; if P_0 is derived from (160) then in a state of rest the surface of the bore will be compressed exactly to its elastic limit.

SHRINKAGE FORMULAS USING PRESSURES AT REST.*

47. As before remarked the relative shrinkage for the state of rest as well as for the state of action is given by (164).

$$\varphi_m = [\,T_m\,] - [\,T'_{m-1}\,]$$

We might apply to this formula the method followed in Article 45 and thus obtain Shrinkage Formulas using Pressures at Rest.

For the *strain* at the inner surface of the outer cylinder we might apply (127) with the notation extended for the particular cylinder considered, substituting for the Pressures in Action $P_1, P_2 \ldots P_m \ldots P_{n-1}$ the corresponding Pressure at Rest $P'_1, P'_2 \ldots P'_m \ldots P'_{n-1}$, obtained as explained in Article 46.

For the *strain* at the outer surface of the inner cylinder we might apply (115), since $P_0 = 0$, with the notation extended, substituting for the Pressures in Action the corresponding Pressures at Rest.

It is however evident that we will obtain Shrinkage Formulas using Pressures at Rest if we make the following substitutions in the formulas of Article 45.

$$\left. \begin{array}{c} P_0 = 0 \\ P_1 = P'_1 \\ P_2 = P'_2 \\ \cdots \cdots \\ \cdots \cdots \\ P_m = P'_m \\ \cdots \cdots \\ \cdots \cdots \\ P_{n-1} = P'_{n-1} \end{array} \right\} \qquad (204)$$

The general equation deduced from (172) will be:

$$\varphi_m = \frac{2P'_m R_m^2 (R_{m+1}^2 - R_0^2)}{E(R_m^2 - R_0^2)(R_{m+1}^2 - R_m^2)} - \frac{2P'_{m+1} R_{m+1}^2}{E(R_{m+1}^2 - R_m^2)} \qquad (205)$$

---

* These formulas are used by the Ordnance Department; U. S. A., in determining shrinkage.

ELEMENTS OF ELASTIC STRENGTH OF GUNS. 103

The Absolute Shrinkage will be:

$$S_m = 2R_m \varphi_m \qquad (206)$$

The absolute contraction of the diameter of the bore, deduced from (115) will be:

$$\Delta_0 = \frac{4R_0 P'_1 R_1^2}{E(R_1^2 - R_0^2)} \qquad (207)$$

The formula for verifying the absolute contraction of the diameter of bore by the partial contractions will be:

$$\Delta_0 = 2R_0 \left\{ \varphi_1 \frac{R_2^2 - R_1^2}{R_2^2 - R_0^2} + \cdot \varphi_m \frac{R_{m+1}^2 - R_m^2}{R_{m+1}^2 - R_0^2} + \cdot \varphi_{n-1} \frac{R_n^2 - R_{n-1}^2}{R_n^2 - R_0^2} \right\} \qquad (208)$$

1° These formulas for a built-up gun of tube and jacket become:

$$\varphi_1 = \frac{2P'_1 R_1^2 (R_2^2 - R_0^2)}{E(R_1^2 - R_0^2)(R_2^2 - R_1^2)} \qquad (209)$$

$$S_1 = 2R_1 \varphi_1 \qquad (210)$$

$$\Delta_0 = 4R_0 \frac{P'_1 R_1^2}{E(R_1^2 - R_0^2)} \qquad (211)$$

2° These formulas for a built-up gun of tube, jacket and hoop become:

$$\varphi_1 = \frac{2P'_1 R_1^2 (R_2^2 - R_0^2)}{E(R_1^2 - R_0^2)(R_2^2 - R_1^2)} - \frac{2P'_2 R_2^2}{E(R_2^2 - R_1^2)} \qquad (212)$$

$$\varphi_2 = \frac{2P'_2 R_2^2 (R_3^2 - R_0^2)}{E(R_2^2 - R_0^2)(R_3^2 - R_2^2)} \qquad (213)$$

$$S_1 = 2R_1 \varphi_1 \qquad (214)$$

$$S_2 = 2R_2\varphi_2 \qquad (215)$$

$$\Delta_0 = 4R_0 \frac{P'_1 R_1^2}{E(R_1^2 - R_0^2)} \qquad (216)$$

$$\Delta_0 = 2R_0 \left\{ \varphi_1 \frac{R_2^2 - R_1^2}{R_2^2 - R_0^2} + \varphi_2 \frac{R_3^2 - R_2^2}{R_3^2 - R_0^2} \right\} \qquad (217)$$

3° These formulas for a built-up **gun of four** cylinders become:

$$\varphi_1 = \frac{2P'_1 R_1^2(R_2^2 - R_0^2)}{E(R_1^2 - R_0^2)(R_2^2 - R_1^2)} - \frac{2P'_2 R_2^2}{E(R_2 - R_1^2)} \qquad (218)$$

$$\varphi_2 = \frac{2P'_2 R_2^2(R_3^2 - R_0^2)}{E(R_2^2 - R_0^2)(R_3^2 - R_2^2)} - \frac{2P'_3 R_3^2}{E(R_3^2 - R_2^2)} \qquad (219)$$

$$\varphi_3 = \frac{2P'_3 R_3^2(R_4^2 - R_0^2)}{E(E_3^2 - R_0^2)(R_4^2 - R_3^2)} \qquad (220)$$

$$S_1 = 2R_1\varphi_1 \qquad (221)$$

$$S_2 = 2R_2\varphi_2 \qquad (222)$$

$$S_3 = 2R_3\varphi_3 \qquad (223)$$

$$\Delta_0 = 4R_0 \frac{P'_1 R_1^2}{E(R_1^2 - R_0^2)} \qquad (224)$$

$$\Delta_0 = 2R_0 \left\{ \varphi_1 \frac{R_2^2 - R_1^2}{R_2^2 - R_0^2} + \varphi_2 \frac{R_3^2 - R_2^2}{R_3^2 - R_0^2} + \varphi_3 \frac{R_4^2 - R_3^2}{R_4^2 - R_0^2} \right\} \qquad (225)$$

APPLICATIONS OF STRAIN FORMULAS.*

48. 1° *5-inch Breech Loading Rifle Siege Gun—Model* 1890. *Section IV—Between shoulder on tube and trunnion-hoop.*

Data:
$$\theta_0 = \rho_0 = 18.75; \; \theta_1 = 21.5$$
$$R_0 = 2.5; \; R_1 = 4.5; \; R_2 = 6.86$$
$$E = 13,393$$

FIRING PRESSURES.

(b, 159), $P_1(\theta) = 7.56$—used in computing shrinkage.
(a, 159), $P_0(\theta) = 18.246$—used in computing shrinkage.
" $P_0(p) = 15.966$—service pressure.
(160) $P_0(\theta) = 22.76$—not used.

SHRINKAGES.

(182) $\varphi_1 = 0''.0012692$
(183) $S_1 = 0''.0114$
(185) $\Delta_0 = 0''.0042$

2°. *8-inch Breech Loading Rifle—Model* 1888. *Section II—Powder Chamber.*

Data:
$$\theta_0 = \rho_0 = 18.5; \; \theta_1 = \rho_1 = 19.5; \; \theta_2 = 21.$$
$$R_0 = 4.75; \; R_1 = 7.5; \; R_2 = 11.7; \; R_3 = 15.$$

FIRING PRESSURES.

(c, 159) $P_2(\theta) = 4.729$—used in computing shrinkage.
(b, 159) $P_1(\theta) = 13.032$—used in computing shrinkage.
" $P_1(p) = $ ——not computed since $P_1 R_1^2 > P_2 R_2^2$
(a, 159) $P_0(\theta) = 23.204$—used in computing shrinkage.
" $P_0(p) = 18.54$—service pressure.
(160) $P_0(\theta) = 23.77$—not used.

---

* The data are taken from "Notes on the Construction of Ordnance—No. 59".

SHRINKAGES.

(186) $\varphi_1 = 0''.001087$
(187) $\varphi_2 = 0''.001259$
(188) $S_1 = 0''.0163$
(189) $S_2 = 0''.0295$
(191) and (192) $\Delta_0 = 0''.0073 + 0''.0052 = 0''.0125$

LONGITUDINAL RESISTANCE.

3°. The breech-block in the 8″ rifle is carried by the jacket and the weakest cross-section of the jacket with respect to longitudinal strain, or transverse rupture, is the annular groove near the front of the breech bushing where the thread in the breech of jacket ends.

Data: $\theta_1 = 19.5$; $R_0 = 4.75$; $R_1 = 7.9$; $R_2 = 11.7$; $P_0 = 23.2$; $P_1 = $; $P_2 = 4.7$.

The longitudinal resistance of this gun is given by (157).

$$L = \frac{3P_0 R_0^2 + 2P_2 R_2^2 - 2P_1 R_1^2}{3(R_2^2 - R_1^2)}$$

The firing pressure P_1 probably extends to the section considered but since P_1 assists the gun in resisting longitudinal strain the most unfavorable case for the gun will be to assume:

$$P_1 = 0$$

we will then have

$$L = \frac{3P_0 R_0^2 + 2P_2 R_2^2}{3(R_2^2 - R_1^2)} = 12.82 \text{ tons}$$

Since $\theta_1 = 19.5$ we see that not quite two-thirds of the longitudinal strength of the jacket is used.

www.ingramcontent.com/pod-product-compliance
Lightning Source LLC
Chambersburg PA
CBHW021944160426
43195CB00011B/1214